Developing Firefighter Resiliency

DEVELOPING FIREFIGHTER RESILIENCY

Bob Carpenter | Dave Gillespie | Ric Jorge

Fire Engineering
BOOKS & VIDEOS

Disclaimer

The recommendations, advice, descriptions, and methods in this book are presented solely for educational purposes. The author and publisher assume no liability whatsoever for any loss or damage that results from the use of any of the material in this book. Use of the material in this book is solely at the risk of the user.

Copyright © 2019 by
PennWell Corporation
1421 South Sheridan Road
Tulsa, Oklahoma 74112-6600 USA

800.752.9764
+1.918.831.9421
sales@pennwell.com
www.FireEngineeringBooks.com

Managing Editor: Mark Haugh
Production Manager: Tony Quinn
Cover Designer: Brandon Ash
Book Designer: Nord Compo

Library of Congress Cataloging-in-Publication Data

Names: Carpenter, Robert, 1963- author. | Gillespie, David, 1968- author. | Jorge, Ric, author.
Title: Developing firefighter resiliency / Bob Carpenter, Dave Gillespie, Ric Jorge.
Description: Tulsa, Oklahoma, USA : PennWell Corporation, [2019] | Includes bibliographical references and index.
Identifiers: LCCN 2018041536 | ISBN 9781593704209
Subjects: LCSH: Fire fighters—Training of—United States. | Fire prevention—United States.
Classification: LCC TH9120 .C37 2019 | DDC 363.37068/3—dc23

All rights reserved. No part of this book may be reproduced, stored in a retrieval system, or transcribed in any form or by any means, electronic or mechanical, including photocopying and recording, without the prior written permission of the publisher.

Printed in the United States of America

1 2 3 4 5   23 22 21 20 19

Contents

Preface xi
    An Unexpected Experience with Valuable Lessons Learned xi
    Notes xiv

PART 1: ADVERSITY OF FIRE

    Introduction 3
1 The Finer Points of Character 5
    The Virtues of Character 7
        Trust 7
        Courage 8
        Bravery 8
        Diligence 9
        Humility 9
        Temperance 9
        Loyalty 9
        Commitment 10
    The Character of Leadership 10
    Notes 11
2 The Anatomy of Fear 13
    Training to Combat Fear 14
    Fear and Age: We Wear Turnout Gear, Not Capes 16
    The Self-Assessment: Are You Listening? 17
    The SNS Trigger 17
    How Stress Originates in the Mind 18
        Sight and sound 19
        Smell and touch 20
    A Voice of Experience—Developing PTSD/Anxiety 21
        Voice of experience—humility 22
        Who can I turn to for help? 22
    Notes 22

PART 2: THE SUCCESS-BASED TRAINING MODEL

Introduction 27

3 The Basics of Adult Learning 29
Successful Steps to Training. 29
    Training vs. drilling. 30
The Five Laws of Adult Learning 30
    Achieving mastery. 31
    Coaching 32
Note. 33

4 The Four *P*s to Success: Plan, Prepare, Present, and Post 35
Plan. 35
    Time. 37
    Logistics. 37
    Location. 37
    Participant knowledge and experience 38
Prepare. 38
    Setting objectives 39
    Practice run of the exercise 41
The Presentation 42
    Instructor to coach 43
    Rehabilitation. 45
Postexercise Debrief. 46
Summary 46
Notes. 47

5 The Safety Plan: Not Just Another Form 49
Features of a Typical Safety Plan. 49
    Drill Objective(s). 50
    Description of Training 50
    Department-related P&P, SOP, and SOG. 50
    PPE/Equipment required 50
    Hazards and Control Measures 51
    Accountability. 51
    In Case of Emergency 52
    Communications 52
    Resources Assigned. 52
    Safety Plan Notes 53
Summary 53
Notes. 58

PART 3: RESILIENCY

Introduction: When It All Comes Together. 61

6 Mental Tactics of High Performance 63
Using the Big Four 65
Taking Control of Yourself. 65
Resiliency 66
Situational Awareness. 66
Mastering the Self. 67
The Three Levels of Self-Awareness. 67
Level I: Basic psychological self-management skills 68
Level II: Intermediate skills 69
Level III: High-performance skills. 69
How Surgeons Practice Skills to Prepare. 71
How Special Forces Practice Skills to Perform 71
Notes. 76

7 Building Mental Toughness 77
Tactic #1: Mental Rehearsal. 79
Types of Mental Rehearsal. 81
Mental imagery. 82
Visualization 83
Mental rehearsal. 84
Three Points of View 85
First-person point of view. 85
Drone point of view 86
Third-person point of view. 88
Developing a Realistic Goal for a Mental Rehearsal Plan 88
Developing a Mental Rehearsal Plan. 89
Example: SCBA donning and activating the PASS 90
Example: SCBA, entanglement, and rescue. 90
Example: SCBA, entanglement, and rescue in greater detail . . . 91
Timing a Mental Rehearsal Plan. 92
Dynamic Imagery 92
Changing Perspective Based on Roles. 95
Repeated Practice 96
Notes. 96

8 The Power of Self-Talk 97
Self-Talk. 97
Thinking Traps. 97
Tactic #2: Constructive Self-Talk. 99

Five Types of Self-Talk in the Fire Service 101
    1. General self-talk of the operation 101
    2. Motivational self-talk...................................... 101
    3. Instructional self-talk..................................... 102
    4. Unrelated self-talk.. 102
    5. Reactionary self-talk...................................... 103
Developing an Anchor Phrase 104
Utilizing Self-Talk as a Proactive Tool............................... 105
    Example of proactive self-talk................................. 105
    Developing an arousal control image or phrase................ 105
Coping Statements 106
    Continued training in self-talk increases confidence 106
Tactic #3: Goal Setting and Segmentation 107
Types of Goals.. 108
    Subjective goals.. 108
    Objective goals... 108
    Outcome goals.. 108
    Performance goals 109
    Process goals 110
SMART Goals... 110
Designing SMART Goals 111
Segmenting a Task When in Crisis.................................. 111
Freezing.. 113
Train, Prepare, and Thrive.. 114
Notes... 114

PART 4: DEVELOPING RESILIENCY THROUGH PHYSIOLOGICAL TACTICS

9 Breathing and Mindfulness... 119
    Belly Breathing (Diaphragmatic Breathing)...................... 119
    The Hum Technique... 120
    Box Breathing.. 121
    Breath and Delivering Air for Safety and Survival............... 122
    The Wheel Technique 122
    Mindfulness.. 124
    Overcoming Negative Self-Talk 125
    Coping Techniques... 126
        Example use of coping statements 127
    Notes.. 128

10	**The Body and Physiological Needs** 129
	Nutrition 129
	Simple nutrition tips 129
	Physical Exercise 130
	Sleep 131
	Notes 133
11	**Training Smarter** 135
	Biometric Tracking 136
	Biometric training example 139
	The Role of Brain Waves on Action and Rest 140
	Mastery 142
	Notes 143
	Index 145
	About the Authors 151

Preface

FEW OF US ARE fortunate enough to find our passion in life. Our careers in the fire service have led to a love affair that has changed our lives. Relationships we have developed over the years with people from across the globe have been an unexpected gift. We've had opportunities to meet our heroes, and some have become our personal mentors. Others have become such close friends that we consider them family. They have inspired us to become fire instructors, and any success we have experienced in the fire service is due to them.

Over the years, we have seen tragedy and horror, but we would be remiss not to mention the joys we also have been blessed to witness. The lives saved, the children birthed, the hearts mended, the people comforted, and the property salvaged has at times made this career seem almost magical. But this book is about a different type of "magic," the type that rescues from despair. The magic identified in this book is *resiliency*.

AN UNEXPECTED EXPERIENCE WITH VALUABLE LESSONS LEARNED

"The only place on that fireground worse than where you were was anyplace else!" These words were spoken by Captain (now Battalion Chief) Kelley Cato upon hearing Captain John Wright recount the events of a hot, windy June night in 2011.

Captain Cato was on the first arriving engine at a residential fire in Flower Mound, Texas. He and his crew were making their way to the command post having exited the structure to get fresh air tanks following a hard-fought initial attack on a stubborn, windblown fire. The fire was consuming a garage and the attached house, which was about 3,500 square feet in size.

The fire resisted extinguishment efforts from the 2½-inch handline deployed by Engine 3. The fire spread to the void space between the garage and the living areas of the home and began to consume the residence. Captain Wright and his crew member, Firefighter Gus Trujillo, expected to be assigned to back up the first engine when they reported to the incident commander. Instead, they were assigned to search the residence. Captain Wright and his firefighter would soon find themselves in a fight for their lives.[1]

> We are told from the beginning of our careers that size-up begins long before the alarm. Weather, time of day, even the season of the year could affect every response during a tour.

Captain Wright recalled thinking about the extreme heat and high winds in the area earlier that day. His experience told him that a fire on this day would challenge crews in conditions some had never experienced before. He was right to be concerned. Listening to the radio transmissions while en route, he began to formulate expectations, recalling fires in similar structures and how they were handled. He communicated these thoughts to his crew, allowing them to benefit from his experience. The process of preparing mentally for an upcoming challenge comes from a state of hypervigilance. Hypervigilance is important to firefighter survival, especially when properly focused through the power of self-talk and visualization. These components of sound fireground actions are discussed throughout this book.

As Captain Wright and Trujillo went to the second floor to search the bedrooms, they could not have known what would happen in the coming moments. During their search they encountered an apparent fire in a void space involving the return air ducts of the home's air conditioning system. Upon opening a wall, a decision they would soon regret, they encountered the fire. Driven by 30 mph winds, the fire began to consume the room they were searching. It was not a big room. Measuring approximately 10 feet by 10 feet, the room usually would have presented no obstacle for the two. Due to the rapid deterioration of conditions, however, they were soon turned around and disoriented. They missed the doorway opening and were cut off from their egress down the interior stairs.

What transpired next exemplified the value of effective training, diligence, and a productive shift in training norms. This training proved life-saving for two capable firefighters, one with an accomplished record and one with little more than what he had learned in recruit training.

The captain immediately radioed command, declaring a Mayday. The initial response was disbelief. Few on the fireground could comprehend how a Mayday condition could have developed in a large but fairly standard residence fire. A quick evaluation of the conditions, however, made it clear that this situation was anything but routine. Command and companies outside shifted their focus from extinguishment to saving their own, facing the likelihood that this might not end well. Others recalled thinking that no one could survive these conditions.

Inside, Captain Wright and Trujillo reverted to their training. The department had been undergoing a comprehensive Mayday/firefighter rescue training process. Beginning with abstract instruction in the classroom and advancing to hands-on scenarios at the training center, the department focused on developing the skill sets needed to survive deadly situations such as this. Even with their training, Wright and Trujillo began to experience some of the physiological and psychological effects that are difficult to replicate, and seldom explored, during training.

Wright recalled that everything sounded muffled and impossible to understand. It was not an equipment problem, but the effects of the sympathetic nervous system (SNS) response under extreme stress. Wright later said he could not really see much. He described having tunnel vision, which is not misplaced focus, but a very real physiological event under extreme stress. Trujillo was able to locate a window and escape the room to the first-floor roof below. Trujillo reached back in through the window and pulled Wright out to the roof. According to one news report, "About seven seconds later, the room flashed."[2] Wright does not recall how he got to the front yard, but he was led there and walked out on his own power. He does remember a moment inside the room when he nearly accepted his fate and was ready to submit, which is the next step in the cascade of responses following fight and flight. Then, just as quickly as things had become a blur, he was able to think clearly again.

Captain Wright and Firefighter Trujillo survived with a few scars to remind them how quickly things can go sideways. Members of the department and many others learned valuable lessons in the months and years that followed. The virtues of temperance, diligence, and courage were readily apparent that night. Disciplined training, as practiced in a success-based format with progressive levels of difficulty, led to a good outcome in a horrible situation. Psychophysical responses such as these men encountered are discussed in the following pages. Understanding these responses and implementing an effective training program is necessary to equip members to survive the hostile environment of the fireground.

Countless incidents like this play out daily in communities throughout the United States. Close calls and line-of-duty deaths (LODDs) are a reality in the fire service. We have long studied these incidents in an attempt to learn what is successful, as well as how to avoid future failures and capitalize on effective performance. In this text, we explain how and why we respond the way we do and provide a process to train our members to perform at the highest possible level.

This book has been inspired by the personal encounters of the authors and is influenced by many. It is the culmination of more than a decade of researching, experimenting with techniques, and speaking with experts in many fields in an attempt to create a training manual that brings together many aspects of firefighting as a whole to enhance the performance of firefighters on and off the job.

NOTES

1. Scott Friedman, "Remembering 'Mayday' Call," NBCDFW.com (September 22, 2011), https://www.nbcdfw.com/news/local/Telling_the_Story_of_Flower_Mound_s__Mayday__Call_Dallas-Fort_Worth-130402638.html.
2. Chris Roark, "First Responders Hear Stories of Heroism on 9/11, Locally," *Lewisville Leader* (October 4, 2011), http://starlocalmedia.com/lewisvilleleader/news/first-responders-hear-stories-of-heroism-on-locally/article_bc227bb9-4b5c-5ab6-b86b-042113db9bfe.html.

PART 1: ADVERSITY OF FIRE

Introduction

Ric Jorge

IT WAS NOT until I began to mature that I could *experience* the value of virtues, and mental comprehension became internal understanding. What I thought about myself became more important than how I thought others perceived me. My maturity came with life experiences, such as the birth of my children. One of the changes that came with maturity was that I developed a much greater capacity for tolerance and understanding. This change was refined in the firehouse when I was trusted to protect my fellow firefighters. Further refinement occurred when I responded to calls where people were at their most vulnerable. I knew that a friendly, humble, and understanding demeanor would have a greater impact than the other more visible successes we would have.

One outcome of the development of compassion was that I realized people trusted me based in part on how I carried myself. The way I carried myself reflected the virtues I internalized to develop my character. How I behaved and reacted when no one else was looking became the barometer with which I measured the growth of my character. This concept may seem simple, but it became vitally important to me. Prior to being diagnosed with post-traumatic stress disorder (PTSD), it was the virtues of my character that kept me fighting to get better. Refusing to give up and striving to get better is called *desire*, and I attribute my successes to the development of desire and other attributes of character.

The purpose of part 1 of this book is to clearly define character and explain the importance of developing a properly focused mind-set. As firefighters learn to develop and control their individual mind-sets, the benefits are multiplied on the fireground and elsewhere. The development of character is critical to the proper mind-set, which is the foundation of resiliency. Character is the keystone to success. The value of virtues when establishing relationships within the fire service, from a servant leadership position and as a "guide" to help develop resiliency in others, is based on Abraham Maslow's hierarchy of needs. The development of your mind-set can benefit the development of the mind-sets of individuals you lead, but only if you understand the *hows* and *whys*.

1

The Finer Points of Character

THE FIRE SERVICE is ruled by type-A personalities who are well-educated, experienced, hard-charging, selfless warriors with can-do attitudes. Yet the practice of developing moral character through mentoring, and the advantages of being a leader with high moral character, seem to have been marginalized over the years.

A discussion of character must first begin with the role of character not only in the fire service at large, but also in the lives of individuals who serve. The qualities and actions that define a person of high moral character extend beyond the fire service, affecting personal lives, as well. There is no escaping the fact that the person you are at home is the person you bring to the fire station, and vice versa. It is not a position or title, but rather character that defines a person.

Having strong character and high moral values helps us avoid problems in the fire service. Character is the key to many solutions and responses, but developing it seems to be a lost art in the fire service. We do not seem to value our greatest commodity: our own people's internal motivation.

The word *moral* in this book does not refer to a pious perspective. Instead, it refers to personal standards or beliefs concerning how people choose to conduct themselves and what behaviors they consider acceptable or unacceptable.

When we look to people as mentors, high moral character is the main quality we seek. Most mentors exhibit leadership qualities such as integrity, fairness, understanding, vision, and inspiration. These individuals are far from perfect; all of us struggle with moral issues in one way or another. In the context of the fire service, however, people of high moral character are true leaders.

Two of the greatest assets in fire service are experience and character. *Experience* is easily identifiable as the amount of exposure or training repetition a person has with a particular subject or skill set. Depending on the extent of that experience, and how it has been applied, it may be viewed as a good or bad experience, or as a practical or impractical experience.

In identifying character, what do you look for? Most people look for particular virtues such as trust, courage, bravery, diligence, humility, temperance, responsibility, loyalty, and commitment. These qualities are important traits when you trust another person with your life.

You must understand character to follow the path of leadership development. Character allows us to understand the logic behind Maslow's hierarchical scale because most people have certain needs and desires. (See fig. 1–1.)

Fig. 1–1. Maslow's hierarchy of needs

You can advance an individual through development of character, and you can advance performance through development of technique. While development of technique does not encourage the development of character, enhancing an individual's character will encourage the enhancement of technique. If a person becomes willing to change, and that willingness is fostered in humility, the door is open to growth and adaptation. Resiliency techniques are effective in part because they help an individual embrace the idea that there is something more to strive for. This is known as *hope*. Humility and hope are virtues of character, and high moral character is the foundation of leadership. Examining the enhancement of an individual or enhancement of a performance model can add clarity to Maslow's hierarchy.

The components of Maslow's hierarchy that focus on safety needs, such as security, law and order, and stability, allow for freedom from fear of physical harm. This continuation of physical needs involves more of a cognitive level than the first step, and gives way to the third step, which is love and a sense of belonging. Establishing friendships, intimacy, affection, and love applies to our families, friends, coworkers, and romantic relationships. Camaraderie can inspire a sense of purpose and motivation.

The fourth step is esteem needs. The ego begins to direct action like mastery, status, self-respect, and autonomy, topics that are addressed in part 2 of this book. One way this step comes to fruition in the fire service is by promotions. The promotional process meets self-esteem needs through the respect given when mastery is achieved.

Step five of Maslow's hierarchy is self-actualization. Do not underestimate the motivation of people who realize their personal potential or experience self-fulfillment. Personal growth can give momentum to continued success and vision.

THE VIRTUES OF CHARACTER

Leadership qualities should be instilled in firefighters from the very beginning because character determines what a person values. Understanding character is vital when educating students. Understanding virtues such as courage and humility, and teaching examples of how character plays an important role in leadership, are elements that are often missing from modern fire service.

It is frequently said that there are two types of firefighters in the fire service: one is a firefighter, and the other just works for a fire department. The first is driven by character, which stokes a passion to be better, to learn, and to strive to achieve progress through a fire service career. The second type of firefighter is there for other reasons, like the pension or insurance benefits, work schedule, pay, and/or ego considerations. It can be argued that this is the attraction behind Hollywood glamorizing fire service careers. Either way, the heart of the lion is missing from the firefighter in the second example.

Virtues such as courage, diligence, humility, temperance, responsibility, loyalty, and commitment are vital leadership qualities. When leaders lead from high moral character, a bond is created, and that unity is imperative to team building. We call it trust.

Trust

Trust is a reliance on someone, or confidence in a system or person. It can even be seen as a dependence on people or a system. In the fire service, trust is paramount to success.

Being tactically sound and skilled in the individual disciplines under your responsibility will bring admiration and trust in your ability (see part 4). The ability to lead, whether as an officer or an informal leader, combined with the ability to explain things clearly and to inspire others, will garner trust. Trust is the cement that binds firefighters. To trust others with your life, and to have others trust you with theirs, is an honor not to be taken lightly.

Courage

Courage is often confused with bravery, but they are very different. Courage is having the mental or moral strength to venture forward and persevere in times of difficulty. At times this is evident in conversations, such as when a courageous leader speaks the truth when others will not. It also can be apparent when leaders face tasks and solve problems or conflicts, whether or not the actions or outcomes are popular.

Example. A revered firefighter began to be regularly late for his shift. Even though there was someone holding over to cover the gap, concerns arose. It also was noted that the firefighter's breath smelled of alcohol. (The smell of alcohol on the breath of an on-duty firefighter should never be overlooked.) When confronted, the firefighter wrote it off as the smell of mouthwash, but the stagger in his gait suggested otherwise. Clearly, an uncomfortable conversation was coming. Often during this type of conversation, concern and offers to provide help are met with contempt and quickly brushed aside. The concerned individual may be told to mind his own business when further attempting to reason with the potentially inebriated firefighter. These are difficult conversations to have, especially in our work place, and it takes courage to have them.

At some point trust enters into the equation, as does respect. A firefighter who fails to perform the simple, yet vital, task of coming in fit for duty must be confronted. Sometimes a leader who has not properly prepared for this type of challenge by developing the necessary character to act prudently will allow this type of behavior to continue. Instead, it should be properly viewed as a liability to the entire company and to everyone's families. In the fire service, it can be difficult to separate our jobs from our personal lives. Looking at the suicide rate of firefighters compared to line-of-duty deaths reveals the side effects and problems this career can bring.[1] We must police our people's mental health for everyone's sake.

Bravery

Bravery is the ability to meet danger or endure pain or hardship without giving way to fear. It is through repetitive training that we develop confidence (see part 2), and this fosters bravery. Bravery is the greatest manager of fear. In simple terms, it is courage put into action. The topic of bravery is further discussed in chapter 2.

Diligence

Diligence can be defined as a "steady, earnest, and energetic effort," and it can make the difference between life and death in the fire service.[2] The careful and meticulous preparation that must be given to fighting fire requires diligence. Training and the ability to concentrate on the details of technique when practicing mark the difference between amateurs and professionals.

There are so many disciplines in the fire service that it takes years of diligent practice to be proficient in all of them. It takes even more years of training and experience to master them.

Humility

Humility can be defined as "freedom from pride or arrogance," or having the honesty to understand that you will never know it all.[3] Asking for help is a sign of strength and courage, not of weakness or lack of education. Character is foundational to developing the ability to face fear.

Temperance

Temperance is the use of restraint that allows for proper appraisal of situations. It is the ability to manage your own thoughts, feelings, and actions with control. Showing restraint in order to properly evaluate a situation or a person, or to follow orders, helps make the difference between a long, successful career and the possibility of a long rehab in a burn unit.

We are often given assignments on a fire scene that, absent the information that the officer giving the assignment has, make no sense. Temperance allows us to place a line in a nonengaged position as ordered and to resist the urge to freelance by moving into an offensive position. When the fire behaves in the manner anticipated by the officer who gave the order, and the line placement stops the advance of fire, the success is not accidental. It is due in part to temperance on behalf of the command officer, who realized the need for the tactic. It is also due to temperance on the part of the fire service members who followed the order that achieved the result. The officer's experiential knowledge provided his temperance. Firefighters must rely on their trust in their commander, use restraint, and follow the order. These virtues do not exist in a vacuum but rather are intertwined throughout our duties.

Loyalty

In the fire service, *loyalty* is devotion to your mission, your partners, and the common welfare of everyone, and it leads to allegiance. Loyalty is developed through repeated actions that are consistent and admirable from a technical or personal perspective. A chief who focuses on meeting the needs of his personnel and their families understands that to lead effectively requires determination to meet those needs. This focus on personnel and their needs earns the loyalty and trust that are vital when bravery and courage are called for.

Commitment

Commitment is an unconditional personal vow to see something through to fruition. It is the willingness to use the full measure of your will and resources to provide the highest level of performance. This is of utmost importance in emergency service work, as mediocrity provides the opportunity for tragedy to occur.

THE CHARACTER OF LEADERSHIP

The virtues we discussed above have deep meaning and demand further contemplation. In the fire service, virtues can be applied in many areas, individually or in large groups, but this does not guarantee the success necessary to lead. When you understand that character shapes leadership traits (vision, motivation, and inspiration), then you can grasp that leadership qualities are not merely benchmarks to ensure success. Character defines an individual, and leading a life of high moral character can easily be described as leading an honorable life.

Great leaders do not sell themselves as "great leaders." They merely live their lives guided by their character. They hold themselves as accountable as they do anyone else. A fire service leader who advances steadily through the ranks has the benefit of understanding the job at a depth necessary to be able to lead successfully from a position of experience. Getting promoted quickly through the ranks, or skipping ranks, can be the sign of someone motivated by ego. Being a "Skippy" is not a badge of honor, no matter how many bugles you wear on your chest. Great leaders do not need bugles to lead. They command respect without asking, and they garner loyalty from their actions rather than from some policy.

Understanding character also allows us to approach the subject of fear, leading us to ask, "Is fear influenced by character?" Fear is a normal human condition, but how we process that response can have a great impact on our lives, and on the lives of our loved ones. If the physiology of fear is not understood, the learning process to overcome anxiety and panic is incomplete. In the fire service, you must know building construction to understand how fire may travel. The knowledge of flow paths allows us to understand why and how fire will develop and travel. Likewise, understanding the physiology of fear allows us to understand the "flow paths" of how anxiety and panic develop and subsequently become manifested in our minds.

Fire science has benefited from ongoing scientific research concerning how the human brain processes information. It has also benefited from the armed services and their use of sports psychology to develop their performance models. Through trial and error and results from scientific and military efforts, much has been learned about how to condition the mind and body. We will make the comparison to the fire service to show how we can better educate and train ourselves and our people to develop and improve mental and physical resiliency.

NOTES

1. "Firefighter/EMS Suicides Outnumbering Line of Duty Deaths," US Patriot Tactical (February 4, 2017), https://blog.uspatriottactical.com/firefighterems-suicides-outnumbering-line-of-duty-deaths/; articles relating to suicide deaths in the fire service, along with prevention efforts, are available at http://www.ffbha.org/resources/press-media/.
2. "Diligence," *Merriam-Webster*, https://www.merriam-webster.com/dictionary/diligence.
3. "Humility," *Merriam-Webster*, https://www.merriam-webster.com/dictionary/humility.

2

The Anatomy of Fear

Ric Jorge

FEAR IS A FOUR-LETTER WORD in the fire service, and it is often associated with weakness. Recognizing fear as a healthy response is the beginning of understanding the sympathetic nervous system (SNS). Fear, directly or indirectly, is one of the reasons we train. Training helps us gain confidence and minimize fear, making us more combat ready. Left unchecked, fear will lead to anxiety, and anxiety will quickly escalate to panic. Panic can be a firefighter's greatest enemy, surpassing even fire in terms of danger. Panic can kill firefighters before fire or smoke can overcome them. The sympathetic nervous system response to stress varies. The mind can be conditioned to embrace extreme responses as normal through training exercises developed purposefully to resist activation of the SNS.[1] It is not unusual for an instructor to observe students experiencing anxiety during confined space training or blackout drills. An instructor should view this response not as a weakness, but as an opportunity for coaching students over an obstacle that could hamper them. This creates a new and healthier response to a stressful situation.

Previous training and real-life experiences will weigh heavily on tactical development. If you were physically restrained as a child, you may feel claustrophobic in some situations. Perhaps you were taunted as a child by an older sibling or friend who would try to frighten you when it was dark. While not everyone may know the exact cause of their anxiety, it is important for instructors and fire officers to understand these dynamics and the possible implications for training outcomes.

When fear occurs during training, regardless of whether the threat is real or perceived, the instructor must recognize and manage the circumstance professionally and knowledgeably. If the threat is real, such as getting burned or trapped, the instructor should take immediate action to ensure the student's safety. If the threat is perceived, such as anxiety or disorientation, the instructor should coach the student toward a successful outcome by helping the student restore rational cognition and behavior. This coaching will lead the student to a better method of handling a similar event should it manifest itself at an emergency scene.

In the fire service, even when we do everything right, the outcome may not be favorable, which is why we train to improve our odds of being successful. The level of anxiety will be inversely proportional to proper training. Lack of proper training leads to greater anxiety when firefighters are faced with circumstances beyond their training and skills. Left unchecked, anxiety will almost certainly develop into panic, and panic renders skills useless. Fear, anxiety, and panic commonly progress in that order, but they may occur in any order based on individual experiences and levels of threat. An astute instructor or officer will recognize when these events are unfolding. This awareness allows for a window of opportunity to intervene and alter the chain of events toward a successful outcome, which instills confidence in the trainee.

TRAINING TO COMBAT FEAR

To associate this concept with a firefighting example, let's use an SCBA confidence course. Usually there is someone on the crew who gets "spooked" during this course. Complaints of phantom regulator malfunctions or mask seal failures are excuses of choice during this drill. Many of us have heard complaints such as, "The air pack was free flowing and would not stop," "I couldn't get a proper face seal," "The regulator was malfunctioning and would not give me enough air," or "My mask failed." After the equipment is inspected, nothing wrong can be found with it. The equipment that needs work in this example is not the hardware, but the "software."

These types of repeated complaints, with no malfunctioning equipment, are clues that you have some work to do with your personnel. Experience dictates that the best way to address this type of problem is not by singling a person out. As instructors, you do not want to lose your student or cause resistance by embarrassment or lack of clarity in direction. In this example, the entire company can drill together to bring this person up to speed (leadership and mentoring done simultaneously), or it can be done individually. Circumstances will dictate which approach is best. Sometimes the officer or instructor can state the need to reinforce teamwork, although the instructor's primary intention is to improve the performance of the student with a problem.

This exercise can be accomplished in the following five-cycle series of three-step exercises.

- **Cycle 1:** To be completed in day uniform
    - Step 1. Everyone goes into the confidence course with all the lights on and doors open.

- Step 2. On the second pass, the doors are closed, but the lights are left on.
- Step 3. The doors are closed, the lights are off, and each person has a flashlight, which is to be used only for orientation. The goal is to build confidence to the point that the light is not used, and the drill can be accomplished in the dark.

- **Cycle 2:** Everyone is wearing their turnout gear.
  - Steps 1, 2, and 3 are repeated as described previously.
- **Cycle 3:** Everyone is wearing turnout gear and SCBA.
  - Steps 1, 2, and 3 are repeated as described previously.
- **Cycle 4:** Everyone is wearing full gear with mask and SCBA.
  - Steps 1, 2, and 3 are repeated as described previously.
- **Cycle 5:** This is the same as cycle 4, except the degree of difficulty begins to increase by adding victims, smoke, sound, strobe lights, and going through the course in reverse direction. These variations are added one at a time, *not* all at once.

The primary goal of this exercise is to help develop your teammates professionally. The secondary goal is to give astute instructors the opportunity to ensure all members of the team can accomplish the first goal. After the instructor correctly diagnoses a problem, he or she must know how to handle the problem. The SCBA confidence course is a confidence course and *not* an attrition course, so use it as one. The idea of building your people up by making them more confident is the foundation of success. Once confidence is established, we can transition our people into developing mastery by increasing the degree of difficulty in a given exercise.[2]

The technique of allowing the student to use the flashlight to briefly reorient oneself while stationary is a subtle way of replacing fear with empowerment. The loss of vision is a very powerful stress trigger. Empowering students by allowing them to use a flashlight at will gives them control over their fear and lessens their anxiety.

This technique shifts the focus away from fear, allowing the student to focus on the task of developing confidence. Another problem an instructor may encounter occurs when a student begins to show signs of trouble during any one of the series. For example, let's say a student shows difficulty between cycles 3 and 4. The plan of action then would be to return the student to the last step in the series in which he or she was successful (competent). The student will work in that area for a while before moving on to the next step or cycle in order to develop the confidence necessary to progress. If the trouble is in cycle 4 only, another option could be to shorten the exercise and allow for shorter successes, with the ultimate goal of creating longer successes over time. This example will hold true for a firefighter being sent to training for remediation in skills that have eroded or where phobias have developed.

This technique is similar at fire academies where students are learning to be firefighters. While at fire department recruit academies (where students are developed to that fire department's specific operations), instructors may utilize a different approach because the recruits are vying for employment in a pass-or-fail environment. It is perfectly understandable to want to hire high-speed, low-drag firefighters rather than trying to "build" or "fix" them.

In each of these scenarios, the instructor will be challenged to clearly identify the student's need and to respond to it as an instructor, coach, or mentor. The training leader must be able to fluidly transition from one role to another as necessary.

FEAR AND AGE: WE WEAR TURNOUT GEAR, NOT CAPES

Some anxieties that can quickly escalate to panic are best overcome with help from mental health clinicians. Post-traumatic stress disorder is a very real side effect of firefighting. For the more easily managed cases, the officer, instructor, or facilitator must recognize that the work being done is more of an inside job than an outside job. It is about the software, not the hardware. Instructors and officers must understand this in order to be effective leaders.

Age can affect performance. What was once attainable is no longer easily attained. Body mass changes with age, and you may not be as strong or thin as you once were. Ego, not character, will convince you to be ashamed of these changes. However, the "secret" that you try to hide might very well lead to your death or the death of a fellow firefighter. This is true for instructors as well as students, and it is the area where character comes into play. An instructor or officer must have moral integrity concerning what is occurring in order to develop a student's weakness into a strength. An instructor or officer with developed leadership qualities can take the unexpected response and come back with innovative brilliance.

Other factors such as encounters with tragic or sensitive calls can alter a person's perspective. Facing your own mortality or fearing for your loved ones, combined with lack of sleep, busy shifts, and physical and mental exhaustion, can leave you with the perfect mixture for a meltdown. How much tragedy can an individual take before he or she begins to break down? This job will take a toll on anyone: year after year of saving lives in impossible situations, cutting cars apart to save trapped passengers, conducting elevated window rescues, saving occupants from being burned alive, responding to collapses, discovering bodies in varying degrees of decomposition, or navigating the countless faces of horrified family members when you tell them their loved one is dead. Those in the fire service get accustomed to being society's problem solvers. We are the last "house" on the block for many, and they measure us by our successes. Our successes make it easy to overlook the fact that we are human, but the reminder will often come from a fatality, a near miss, an LODD, or other cumulative factors.

THE SELF-ASSESSMENT: ARE YOU LISTENING?

Do you have the courage to make an honest self-assessment? Are you humble enough to admit when things are "not right" in order to get the help you need? These issues are why the fire service should support mentoring to develop character within our ranks. In struggles like these, character will help carry you through. Personal recovery depends on the courage to conduct an honest self-examination of your life. When you need help, the ability to trust and the humility to reach out are good foundations for taking the first step. The commitment necessary to mend our fractures will require the utmost diligence. Being virtuous enough to navigate through life is recognized as leading a life of honor. It is not a new concept. From the history of Sparta to the modern military, people of honor have demonstrated great leadership. Character has always been the path to finding our way home. To do an honest self-assessment of where you are psychologically, mentally, emotionally, and physically requires the courage to face your own fears. To be an effective instructor, fire officer, and leader, you must practice what you preach.

THE SNS TRIGGER

The sympathetic nervous system (SNS) is an evolutionary tool that has helped man to survive throughout history. When faced with a real or perceived threat, the heart will begin to speed up, respirations will increase, palms may get sweaty, peripheral vision and hearing may be altered.

A range of situations can cause this reaction, including actual life-threatening situations such as falling through a roof or floor, or being caught in a collapse. Other less dangerous (but still stressful) situations can also cause this reaction, such as the process of promotional interviews, working up the courage to ask that special someone to marry you, or the possibility of facing the tax attorneys for the IRS. All that is needed is a threat (real or perceived) to conjure the response associated with fear, and the heart will beat faster and respirations will increase.

The release of a host of hormones will cause a broad range of reactions. Stress is a very powerful foe that takes its toll on the body over long periods of time, much like a slow cancer. If left untreated, stress can have just as serious health effects. The US Fire Administration (USFA) has provided evidence through National Institute for Occupational Safety and Health (NIOSH) reports of LODDs that fit the criteria of the sympathetic nervous system response leading to panic. This occurs when well-trained professionals behave in uncharacteristic ways, leading to disaster. At least three NIOSH reports (F2011/18, F2007/18, and F2001/13) clearly show what is being described here.[3]

Firefighters in the NIOSH reports exhibited behavior consistent with the SNS response. Erratic behavior from a well-trained and conditioned person should be an immediate indicator that something is wrong.

The truth of the matter is there are many reports like these that can be found in the archives of the USFA library of NIOSH reports involving LODDs. This type of occurrence may not have received much attention for several reasons. Sometimes the documentation has not been classified consistently. Also, for many years little was known about the SNS response in firefighting, and what was known was not understood well enough to make the association and offer recognized solutions. Sometimes the SNS response will be triggered under less-threatening conditions. Regardless of the trigger, stress is always the underlying motivator. Stress activates the SNS response, which will alter your perspective, your self-control, and your control over a situation.

HOW STRESS ORIGINATES IN THE MIND

Stress is primarily triggered through the senses in the form of fear, anxiety, and panic. Memory is formed as raw information from the senses and the emotion is processed by the hippocampus and the amygdala. The amygdala is involved with consolidating or processing memories with greater emotional involvement, especially related to fear. As the emotional intensity of memories increases, so does the possibility of the memories becoming fragmented or distorted. This can explain why sometimes two people will have very different memories of a shared experience.

Processing of memory by the brain on both a structural and biochemical level is very complex. However, in simplest terms, the formation of memory can be divided into two groups based on the anatomic pathways (hippocampus vs. amygdala) of the sensory processing:

- Group 1: Sight and sound
- Group 2: Smell and touch

Sight and sound

In the first group, sight or sound travels by way of the thalamus, which breaks down visual cues by size, shape, and color, to the cerebral cortex. The cortex allows us to give meaning to sights or sounds, enabling us to be conscious of what we see or hear.

In contrast, the second group (smell or touch) is processed directly by the amygdala, bypassing the thalamus. If the fear response is triggered, it is only after the fear response kicks in that the conscious mind recognizes it. When the threat has passed, the anxiety response typically is turned off. It may not, however, if it is influenced by hypervigilence or PTSD.

The triggering process is the main difference between the two groupings of the senses (understanding this is vital when creating training exercises). When the amygdala is stimulated and the fear response is triggered, the bed nucleus of the stria terminalis (BNST, which is sometimes referred to as the *extended amygdala*) can extend the duration of fear states producing the classic signs of anxiety, including rapid heart rate, increased blood pressure and respirations, hyperalertness, and goose bumps.[4]

The hypothalamus acts as the mediator between the nervous system and the endocrine system by releasing hormones that drive the production of cortisol. "Acute stress causes rapid release of norepinephrine (NE) and glucocorticoids (GCs), both of which bind to hippocampal receptors. This release continues, at varying concentrations, for several hours following the stressful event, and has powerful effects on hippocampally-dependent memory that generally promote acquisition and consolidation while impairing retrieval."[5]

In simple terms, the stress hormones cause the cells of the hippocampus to short circuit. The hippocampus is the keeper of long-term and short-term memory, and it also facilitates spatial navigation. So, the cortisol effect on the hippocampus can result in the memories of traumatic or stressful events becoming fragmented, causing them to lose proper context. Simultaneously, adrenalin floods the muscles to prepare for fight, flight, or freeze. The combination of stress, hormonal influences, and adrenaline can lead to uncoordinated movements, such as an inability to swing an axe appropriately or to walk to a location as instructed.

Two of the senses, one out of each group, can be classified as more influential than the others. Vision is the predominant sense, so much so that it can override other senses. Vision is so strong it can override our sense of hearing. How we all use visual speech information and integrate it into what we hear is called the *McGurk effect*.[6]

Smell and touch

The sense of smell, on the other hand, is so powerful that it can trigger memories with a strong emotional component before the mind is conscious of a response. It can be a positive memory response or a negative memory response, depending on the smell. For example, the smell of hot apple pie may bring back fond childhood memories, triggering a positive response. However, if you have a severe allergy to apples, and you had bad reactions as a child, the smell might bring back negative memories and trigger a negative response. The smell goes directly to the amygdala and generates an immediate associated response, in the same manner that unknowingly placing your hand on something hot will cause you to jerk your hand away. All of this occurs before the mind can be made aware of what is going on. The term *making sense* of something takes on a whole new meaning after we understand how the senses affect our thought processing.

Victims of PTSD and other such disorders often describe that they feel as if they are "losing their minds." If we understand the physiology of how the brain processes fear or trauma, it makes sense that these memories could be fragmented or scattered. Many people need professional help when it comes to PTSD because overcoming it could require therapy or medication, or both.

One form of treatment for PTSD is called *exposure therapy*. This is a process where a trained professional takes the victim through a detailed process of recounting an event or series of events. The incorporation of all the senses to create as vivid a memory as possible of a situation is vital. It allows for proper thought restructuring. Using the apple pie example, given previously under the sense of smell, the scenario would be recreated so the victim can understand that apple pie is not bad, but eating it is. The association with apple pie in this example can bring back traumatic memories to the point that even the smell of apple pie could be difficult. This example can be applied to people who have fears of flying, darkness, falling, water, confined spaces, and many other examples. The symptoms of fear have the potential to develop into anxiety, and ultimately panic. Understanding some basic mechanics of how the brain works helps to explain some human reactions in certain situations. These initial fears began somewhere, and understanding the physiology of how these issues begin allows for the perspective that these are normal reactions to abnormal circumstances. All this information is significant to a good instructor. A fire service instructor should understand the mechanics of how and why the brain processes fear and trauma to be able to recognize symptoms of anxiety and panic as possibly something more significant than just fear. A fear might be based on a lack of familiarity with a certain discipline, such as blackout work during SCBA confidence drill. The difference between a natural fear response and a fear response stemming from a previous trauma may become evident by whether or not the student progresses.

A VOICE OF EXPERIENCE—DEVELOPING PTSD/ANXIETY

As my skills as a fireman became more advanced, I fancied myself a high-speed, low-drag fireman with a can-do attitude. I was not prepared in the least for what the accumulation of a career of tragedy can do to a person. In fact, I thought I had dealt with many of the tragedies I had encountered. I had excellent support groups within and without the fire service. My family was supportive, and I have been blessed with good health. So, what could go wrong?

In reality, what I was going through was nothing short of emasculating, and there was not much information readily available. I am not the type of man who will shy away from confrontation. I have a long history of coming straight at my problems (I have a typical firefighter type-A personality), but I was not prepared for the problem I was experiencing. I began to feel weak, as if I was losing control of my life and circumstances I once dominated. I became hypervigilant, expecting a threat almost everywhere I went. My sleep became erratic, and the nightmares were constant. My thoughts were jumbled and focus was next to impossible, but the anxiety was what really pushed me over the edge. The anxiety of not being able to breathe was consuming me. Something as innocent as lying in bed with a sheet over my face would trigger my anxiety. I could not tolerate it. I felt as if I were suffocating. It was not just the need to get out of bed, but to bolt out of bed, and walk (almost run) around my home as if I were being stalked. I had become so overwhelmingly emotional that the only way to keep people from recognizing something was wrong was to rage. So long as my anger could keep people at bay, my secret was safe.

Having been a firefighter for decades worked against me because I thought I should be able to handle anything. I thought that because I had encountered more than the average person, I should be better skilled at handling my "issues." A personal tragedy was the vehicle for my diagnosis of post-traumatic stress disorder (PTSD). My diagnosis was the beginning of a project I never saw a need for. Ironically, PTSD was the missing piece. It was in my brokenness that I found peace and was made whole.

Voice of experience—humility

After more than 20 years in the fire service, I suddenly found myself with a claustrophobia that was so encompassing I would at times want to rip off my SCBA mask. My breathing would become erratic, and my thoughts wild. Why was this happening to me?

Who can I turn to for help?

An employee assistance program was my saving grace. I learned to work through fear by facing it using tools I did not know existed. I learned to recognize the patterns that were detrimental to my mental health. It was this recognition that would allow me to use my new-found tools to defeat my fear.

NOTES

1. Brent Robbins and Harris Friedman, "Resiliency as a Virtue: Contributions from Humanistic and Positive Psychology," in *Continuity versus Creative Response to Challenge: The Primacy of Resilience and Resourcefulness in Life and Therapy*, eds. M. J. Celinski and K. M. Gow (New York: Nova, 2011). Chapter accessible at www.academia.edu/281393/Resiliency_as_a_Virtue_Contributions_from _Humanistic_and_Positive_Psychology.
2. "Yerkes-Dodson Law," *Wikipedia*, https://en.wikipedia.org/wiki/Yerkes%E2%80%93Dodson_law.
3. NIOSH, NIOSH Report No. F2011-18: Death in the Line of Duty . . . A Summary of a NIOSH Fire Fighter Fatality Investigation: A Career Captain Dies and 9 Fire Fighters Injured in a Multistory Medical Building Fire—North Carolina, FACE Report F2011-18 (Morgantown, WV: US Dept. of Health and Human Services, Centers for Disease Control and Prevention, National Institute for Occupational Safety and Health, 2012), https://www.cdc.gov/niosh/fire/pdfs/face201118.pdf; NIOSH, NIOSH Report No. F2007-18: Death in the Line of Duty . . . A Summary of a NIOSH Fire Fighter Fatality Investigation: Nine Career Fire Fighters Die in Rapid Fire Progression at Commercial Furniture Showroom—South Carolina, FACE Report F2007-18 (Morgantown, WV: US Dept. of Health and Human Services, Centers for Disease Control and Prevention, National Institute for Occupational Safety and Health, 2009), https://www.cdc.gov/niosh/fire/pdfs/face200718.pdf; NIOSH, NIOSH Report No. F2001-13: Death in the Line of Duty . . . A Summary of a NIOSH Fire Fighter Fatality Investigation: Supermarket Fire Claims the Life of One Career Fire Fighter and Critically Injures Another Career Fire Fighter—Arizona, FACE Report F2001-13 (Morgantown, WV: US Dept. of Health and Human Services, Centers for Disease Control and Prevention, National Institute for Occupational Safety and Health, 2002), https://www.cdc.gov/niosh/fire/reports/face200113.html.
4. M. A. Lebow and A. Chen, "Overshadowed by the Amygdala: The Bed Nuclease of the Stria Terminalis Emerges as Key to Psychiatric Disorders," *Molecular Psychiatry* (April 2016), https://www.ncbi.nlm.nih.gov/pubmed/26878891.

5. Danielle M. Osborne, Jiah Pearson-Leary, and Ewan C. McNay, "The Neuroenergetics of Stress Hormones in the Hippocampus and Implications for Memory," *Frontiers in Neuroscience* (May 6, 2015), https://doi.org/10.3389/fnins.2015.00164.
6. Harry McGurk and John MacDonald, "Hearing Lips and Seeing Voices," *Nature* 264 (December 23, 1976): 746–748, doi: 10.1038/264746a0.

PART 2:
THE SUCCESS-BASED TRAINING MODEL

Introduction

Bob Carpenter

I **FOUND MYSELF** assigned to the training division midway through my 30-year tenure in the fire service. A department-wide training initiative was formed based on the Department of Defense's curriculum concerning response to terrorism and weapons of mass destruction. The program was well laid out in module form for classroom presentation and concluded with a brief field exercise setting up the decontamination corridor. When I was assigned to teach a module one day, I balked. How was I supposed to present a class on a subject that I knew practically nothing about? I explained that I was in no way an expert on weapons of mass destruction (WMD). My boss's response was, "That's okay. You're on Tuesday." This began my attempt to learn how to teach.

The state-mandated class to become a fire service instructor had not really prepared me to teach. Some theory was discussed in the textbook, but what lacked was Maslow's hierarchy and how that information can be used to teach more effectively. The curriculum had changed since the first time I got certified. The new required reading had improved, but I still struggled with how to get this from my head to my hands. I watched the other trainers. I watched the attendees even more closely. Body language, participation, engagement, and verbal cues began to make sense and bring it together for me.

Few attendees had a real personal interest in this mandated training. The initial PowerPoint slides did little to draw in the audience. There were a ridiculous number of "Course Objective" slides that read, "After this class, the participant will be able to…." These slides detailed what the students would be able to accomplish, understand, or demonstrate after the class. Had those assertions been true, this would not have been an eight-hour class. Instead, it would have been several weeks long. The curriculum was required, however, and there was no option to alter the program. So I studied the material, worked on my presentation at home, and did the best I could.

Trusted friend Captain Michael Posner visited me a few years later and talked me into returning to the Training Division, where I took an open position in Operations Training. Captain Elvin Gonzalez rounded out the countywide team, developing training initiatives and "taking the show on the road." Since we lacked a formal training center at the time, acquired structures, building sites, and vacant parking lots provided the backdrop. Soon we were getting calls from company officers, asking about how they could prepare company drills with similar participation and enthusiasm. Our successful efforts evolved to larger-scale events.

We started to take a hard look at what we were doing differently from those who came before us. No criticism was directed at any of the training officers who preceded us. Some of the training officers before us followed the institutionalized format they were subjected to because they lacked any other options. Most, if not all, instructors were as overwhelmed as they were understaffed, underfunded, and underappreciated. Some did not really want to be there, but as new junior officers, that was where they were assigned. Sometimes their success was limited because their interest in the position was somewhat limited. Training successes are person-driven and not position-driven.

3

The Basics of Adult Learning

THE PSYCHOLOGY OF LEARNING is complex and will be examined as it relates to techniques that help people learn. We have shared and used the following techniques in many forums. If we understand *how* and *why* adults learn, then we can understand why some methods work and others are doomed to failure. Coupled with the psychophysical aspects discussed in parts 1 and 3 of this text, it should become clear that developing a success-based training model is essential.

We will address a purposeful, organized approach to follow, enabling the trainer to move from concept to application in the development of company-level and larger training events.

Finally, we review how to use a formal safety plan (chapter 5), which is not an effort to "safety us to death." Rather, it is a comprehensive document that helps the instructor mitigate potential hazards on the training ground. It also serves as a planning template to help follow an organized approach.

SUCCESSFUL STEPS TO TRAINING

Let's examine successful steps to designing training exercises. An organized approach from concept to delivery will help keep the instructor on track and maximize efficiency on training day. Much has been written about the psychology of learning, and there is no shortage of choices for required reading on the path to becoming a fire service instructor or trainer. Too often, little time is spent to ensure that instructors understand the science in practice. Consequently, when someone is appointed, selected, elected, or promoted to the training officer's post, they mimic what they experienced under their own training officers, for better or worse. If they had more bad examples than good, they tend to perpetuate a dysfunctional cycle. Morale suffers, attendance wanes, and most importantly, learning simply does not take place. Conversely, with a solid understanding of adult learning needs and constructive teaching styles, an enthusiastic training culture can thrive.

Training vs. drilling

Training and *drilling* are terms we use synonymously, but in practical application, they are distinctly different disciplines. When I think of training, I think "beginning" (competency), and when I think of drilling, I think "perfecting" (mastery). The adult mind needs association or relevance to make the lesson meaningful, let alone memorable. When new information or skills are introduced, we instinctively try to match them with information we already know, combining them to form a new understanding or skill. As mentioned in the previous discussion of Maslow's theory, the adult learner needs to associate the lesson or lessons with satisfying basic human needs. Nothing is more basic for adults than being able to provide food, clothing, and shelter for themselves and their families. Therefore, the instructor must establish the connection between training and drilling and the ability to do a better or safer job, thus meeting these basic personal and family needs. If there is no connection, there is no attention.

The primary difference between training and drilling is the level of performance or terminal object being sought. If we agree on that distinction, it stands that training seeks a level of competency, while drilling seeks a level of mastery. Thus the methods and activities to achieve those levels should necessarily be different. Understanding this concept is the first step to success-based training.

THE FIVE LAWS OF ADULT LEARNING

There are many theories about the psychology of learning. The five laws of learning often discussed are as follows:

- The law of readiness
- The law of primacy
- The law of repetition
- The law of intensity
- The law of recency

Let's address how these apply to our training. The law of readiness simply states that the brain needs to be prepared to receive the information. To that end, as we understand that the adult mind learns best when the basic human needs are addressed, it must be clear that the lesson or information is relevant. The most readily received lessons for adults are usually job related. The adult mind often validates new information or processes it based on known information. Combining new data or skills with previous knowledge results in new information or processes. The new skill achieved in training should be measurable as to competency. A certain amount of repetition usually results in a reliable baseline performance. This, however, is a far cry from mastery.

Some folks are what I like to call professional degree seekers, as they love learning new stuff whether they will use the information or not. That is not the norm, however. For example, imagine trying to learn a new language just for the sake of learning it. How likely is your success? Now, if the only way for you to succeed or survive the job market would be to speak that language, your success is much more likely. The most basic human need of security would be met, so the readiness is established. Add the possibility of job termination if you cannot communicate, and the law of primacy is addressed.

The law of primacy can have a dual meaning. Merriam-Webster defines *primacy* as "the state of being first (as in importance, order, or rank)."[1] Here, it means that the method learned first (primary) is the method first recalled. Also, what is before you is of primary importance and deserves your undivided attention. These two laws must be addressed on the path to both terminal objectives—competency and mastery. Once competency is established, the readiness and primacy are in place, usually needing only a reminder, perhaps in the form of a review, before moving on to master the skill. The importance of establishing solid, competent understanding or performance cannot be overstated. Competence begets confidence, and confidence begets competence. It is immaterial which comes first because the two coexist interdependently and perpetually. Competence and confidence are the building blocks for mastery. Through repetition competence is embedded.

Achieving mastery

Confidence in a skill does not a master make. Remember the old saying, "Practice makes perfect"? That is not necessarily the case. Practice does not make perfect; perfect practice makes perfect. This is done by activating the next three laws of learning: repetition, intensity, and recency. Previously, we discussed repetition to achieve competence. Repetition, along with intensity in the form of related tasks that are progressively more difficult, will bring the learner to the point of muscle and motor memory. People who have mastered a skill do not just practice until they get it right, but until they cannot get it wrong. The law of recency applies twofold. The most recently practiced skill is the most easily recalled, and recency in the form of revisiting skill practice provides maintenance of mastery.

Perhaps the biggest impediment to firefighter learning is producing training events or drills that require expert or mastery level performance without first establishing competency or the basics. Think back to your early training experiences. What was the first search and rescue drill like? It probably went something like this: 300+ feet of hose twisted, intertwined, and tangled in the truck bay traversing numerous obstacles like pallets, furniture, or perhaps even the undercarriage of the apparatus. Your vision was probably blacked out by some means, or vision-obscuring Hollywood smoke may have been pumped into the room. You may have been told to find the end of the hose, and you began searching the room. Somewhere along the line, quite predictably, you passed the point of no return and began to run low on air and then eventually ran out of air. You were probably instructed to continue "dirty breathing." You never found the victim, and you finally conceded defeat. It may have ended with a dressing down that would make a drill sergeant blush. Does this sound familiar? This is an example of failure-based training in that it would require a seasoned, skilled firefighter to maximize forward progress and minimize time and air consumption in order to achieve the objective. Furthermore, it enforces foolish and dangerous practices like ignoring air consumption and warning signals, while expecting members to later practice sound air management on the fireground. Yet in firehouses all over the country, this exercise is played out in the name of training.

The above scenario is anything but training. It has the element of intensity that would move one toward mastery, but absent strong basic search skills, it is little more than a test, and a bad one at that. There is no expectation of success for the participant by either the instructor or the participant. In fact, some instructors are known to change the objectives if it appears that the participant is on the verge of success. It is almost as if they feel that student achievement of the objective somehow reflects badly on the instructor or invalidates the exercise.

Coaching

Learning events, if they are to be true learning events, *must* be designed with an expectation of success. Such events should allow for a possibility of failure, rather than have an expectation of failure with only a possibility of success. The latter is illustrated above. Any failure of performance is the opportunity to coach and guide. That is where the real learning of the complex actions under stressful conditions happens. Difficult, challenging objectives are fine. Impossible objectives are not. Participants should not be left to struggle to the point of total frustration and ultimate failure. With basic skills firmly in place, allowing firefighters to work through a problem is very effective in implanting the solution in their brains for recall under stress. If necessary, coaching to get them there accomplishes the same end. The bonus is that the instructor will also earn student trust, and it is through trust, not intimidation, that the instructor earns respect. Under high-stress events we do not rise to the occasion, but we fall back to our level of training.

If a firefighter's last experience executing a skill ended in abject failure under stress on the training ground, do not expect different results under the stress of the fireground.

We know that there is a distinct difference between training and drilling. This brings us to a third term, which is *testing*. Testing serves two main purposes. It can be used as a needs assessment to confirm basic skills to determine future training content, and it can also be used to measure the success of delivered curricula. Drill day or training night often is more of a test than the introduction of new material and skills or the opportunity to repeat and intensify a skill activity. This is especially the case if the activity is not within a time frame that would qualify as recent training.

Because this type of event has merit, departments should develop "no-notice" drills as periodic needs assessments. The no-notice event may also serve as a follow-up for assigned task development. For example, some departments regularly use *NFPA 1410: Standard on Training for Emergency Scene Operations* as a drill-of-the-month activity, with the training division officers visiting the stations at random to spot check compliance and conformance. Sometimes an exercise serves to expose skill weakness rather than to confirm skill prowess. When this is the case, the outcome must be utilized to move the training program forward constructively, not punitively. Based on the results, instructors and trainers should formulate a plan to rebuild or instill competency, moving forward with activities that allow for repetition and increased intensity to achieve mastery. Revisiting those skills, especially the infrequently used ones, serves to maintain high performance expectations.

Whether served by volunteer, paid-on-call, or career departments, the public expects high-caliber performance from its fire companies. With recruitment and retention challenges in many organizations, we cannot afford to run off willing civil servants by washing out trainees with failed training strategies. This job is not for everyone, and not everyone can or should make the cut. As trainers and instructors, we are obliged to provide the information and opportunity for them to achieve success.

NOTE

1. "Primacy," *Merriam-Webster.com*, https://www.merriam-webster.com/dictionary/primacy.

4

The Four *P*s to Success: Plan, Prepare, Present, and Post

Bob Carpenter

TWO KEYS to a successful training event are organization and planning. While it is true that one can take advantage of a learning moment that presents itself, preparing for a lesson ahead of time usually pays the highest dividends. Let us look at four distinctive steps to accomplish this goal.

PLAN

The initial phase involves selecting the topic. This needs careful consideration. Often, the impetus to train stems from mistakes made on actual calls or from perceived shortcomings, and this is certainly important. There is something to learn and improve upon on virtually every incident. However, if the stimulus to train is routinely harping on mistakes, the members can begin to feel like training is punishment. The fire service is rich in stories about the "all-night drill" after a subpar performance at an emergency scene. While a refresher on proper execution is warranted, officers and instructors should be careful to avoid creating a barrier to learning. Remediation should always be constructive, not punitive. The goal is to improve performance and ensure competency. The all-night drill may not change the performance of the member in question, and the rest of the crew often suffers for the effort. Further, supplanting labor-intensive, repetitive evolutions for general discipline issues is counterproductive. The ne'er-do-well is not going to do a better job on station duties or come to work on time because he or she hates doing forward lay evolutions. There are plenty of undesirable details around the firehouse. Employing better leadership techniques, coaching, and formal discipline are methods better suited for correcting work habits.

If the topic for the exercise is selected due to specific poor performance on a run, focus on that deficiency directly and constructively. Demonstrate the process and provide an opportunity for the member to understand and execute tasks. Follow up later with a detailed learning exercise to gain or restore competency and revisit to move the person toward mastery.

Often overlooked as a reason for review is proper execution or performance. When members walk away from a call proud of the outcome or are praised, especially by other companies, take the opportunity to review why the call went well. Emphasize specifically what actions contributed to the success. If the actions were that of another crew, enlist those members to host training on the topic at a future opportunity. Doing so acknowledges the efforts of those who keep their skills honed to a sharp edge and provides for peer leadership. Subordinates are sometimes more responsive to instruction from peers they respect.

When choosing the topic of the exercise, the focus of the activity will affect every aspect of the process, from logistics to number of participants needed or available to the time required or available. Single-company exercises focus on individual skill development, company task development, and position rotation and cross training within the company (fig. 4–1). Conversely, the multi-company exercise is better suited for most scenario-based exercises. Multi-company activities are also suited for concurrent company operations and strategic application of company-level tactics.

Focus of Activity

Single-Company

- Individual skill development
- Single company task development
- Position rotation through skills / evolutions

Multi-Company

- Larger scenario based exercise
- Emphasis on concurrent operations
- Strategic application of company level tactics

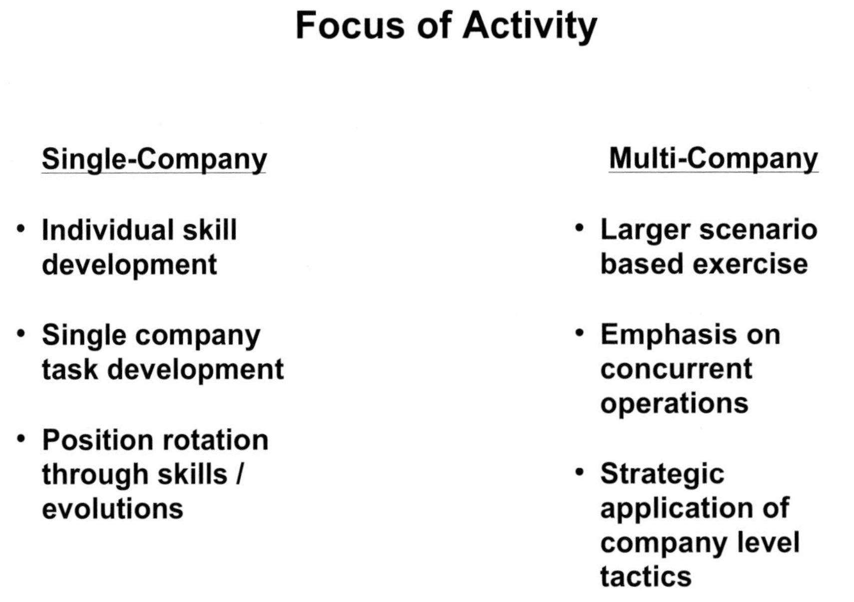

Fig. 4–1. Consider the focus of activity based on a single or multicompany audience.

The National Fire Protection Association (NFPA) provides a good baseline for company evolutions in *NFPA 1410: Standard for Training on Emergency Scene Operations*. Such evolutions are easily adaptable to local staffing conditions and specific equipment and appliances. Miami-Dade (FL) utilized these evolutions as a department-wide competition in 2009. This measure served as a great morale booster for members. It also sparked innovation by the members, resulting in reconfiguration of some apparatus to allow more efficient access and deployment of appliances and equipment. There are evolutions included in the standard applicable to single-company as well as multicompany operations.

Time

The time available or allotted for the session must be considered, along with the topic. Whether the company is in service and available for calls or out of service certainly will affect the scope of the exercise. If out of service, arrangement for coverage must be addressed.

Logistics

Complex exercises can have a lot of moving parts. Again, utilizing units that are in service and available for calls can sidetrack the best intentions of the trainer. Therefore, acquiring tools, appliances, and other equipment that may be used and deployed without removing items from the response vehicle is a plus. This is not the time to press into service old, obsolete equipment. Equipment used in training must be the same type and operation as those utilized in response mode. For example, utilizing an older apparatus that has 3-inch supply hose for hose evolutions is of little use if your department utilizes large diameter hose (LDH) for supply lines. Sexless couplings, hydrant assist valves, water thieves, and spanners are completely different than their counterparts for conventional hose and appliances. If, however, your department uses both, then exercises should be conducted equally to ensure consistency.

Rapid intervention crew (RIC) training can be very time-consuming due to the need to reset and reload RIC bags and equipment as teams conduct individual exercises. If extra equipment is at the ready at the conclusion of an evolution, the next team can begin, and the previous crew or assistants can reload the used equipment.

Location

Whether utilizing a fixed fire department training facility or an acquired structure or a vacant lot or field, confirmation of access needs to occur ahead of time. If the site is not fire department property, proper authorization must be secured in advance. Previous use of a site for training activities does not constitute permission for additional use in the future. It is also important to verify that the party granting permission has the authority to do so. Remember when using an "acquired" structure that *vacant* does not mean *abandoned*. All property belongs to someone.

Logistical considerations include prudent safety measures. Operations that include gasoline-powered saws and other machinery should have, at a minimum, portable extinguishers at the ready, but not mounted on the apparatus. Caution is needed when refueling and cutting, particularly when using rotary saws, which can produce sparks. This is dangerous in an area where saws may have leaked fuel, or fuel may have been spilled while hastily refilling. Around the country firefighters have been burned when fuel was either spilled on the ground or leaked onto the operator's gear and then was ignited by the sparks produced in the operation. Readily available extinguishers may reduce injuries or the extent of damage.

The fire service is a dangerous vocation, and what we do is hazardous. What we do in preparation for service is also dangerous. That does not mean, however, that we should expose our members to undue hazards on the training ground. On the contrary, we have the luxury of forethought and time to mitigate much of that danger on the training ground. (See Safety Plan in chapter 5.)

Participant knowledge and experience

A participant's experience and existing knowledge is important early in the development process. Minimal existing skill sets and inexperience will affect time required, scope of operations, and physical resources. Be cautious when relying on a member's longevity as an indication of experience. Just as we have all met the 2-year veteran with 25 years worth of opinions, there are plenty of 20-year veterans with 2 years worth of skills.

PREPARE

Preparation is separate from the initial planning phase. In the preparation phase, we determine the objectives based on the elements previously discussed. This step includes doing research, if necessary, to provide supporting evidence for the conclusions based on the needs assessment or to support a change of process or a completely new procedure. Map out the actual design of the exercise and secure needed resources and training aids based on previously determined components.

Setting objectives

Setting objectives that are achievable is at the very heart of success-based training. Far too often, training officers approach this component with the wrong mentality. We have been conditioned to believe that if the exercise is not difficult or nearly impossible, it reflects badly on the instructor. This simply is not true. In fact, the very opposite is more likely. Exercises designed with an expectation of failure, whether the expectation is on the part of the participant or the trainer, are likely to be self-fulfilling prophecies. This is not to say that the activity must always be easy.

Fig. 4–2. In fire training, we are not in the business of handing out "Also Played" trophies. We *are* in the business of preparing ordinary people to do extraordinary things, and this does not happen by accident.

Solid objectives are clear, concise, measurable, achievable, and repeatable. Some evolutions have predetermined standards, such as with the *NFPA 1410: Standard for Training on Emergency Scene Operations*. The time for execution is set, the specific components of the evolution are stated, and the standard has been validated. Similarly, performance objectives in many state certification processes enumerate these same components but also identify what actions or inactions constitute failure or unacceptable completion.

> Activities should be designed with an expectation of success with the *possibility* of failure, rather than an expectation of failure with the *possibility* of success.

Remember that it is okay for the activity to be difficult, but not for the sake of being difficult. To that end, it is incumbent on the trainer to provide the information necessary to succeed. This can be accomplished many ways, depending on the subject and complexity of the process at hand. It may come in the form of advanced reading before the training event, a classroom session reviewing the information, or a simple hands-on demonstration prior to skill execution. This process is not "spoon-feeding," but teaching. Everything else is just testing. Having members report to a drill site for a secret, no-notice drill is not a drill. It is a test, and that is acceptable if you are seeking one of the two things that a test provides. A test determines a baseline of knowledge or skill and provides a needs assessment. This is very useful if the result of the observed performance guides the development of future learning activities. A test also confirms that the transfer of knowledge has occurred at the conclusion of a block of instruction and that terminal objectives of competence or mastery have been met. Conducting what amounts to a never-ending string of tests does little to advance the capability of your team members. As discussed previously, training seeks or provides competency, while drilling (law of intensity and law of repetition) seeks or provides mastery.

Ensure that participants know what is expected of them. Provide them with the information to meet the objectives and guide them to do just that. Members need to trust their trainers if they expect to get out of their comfort zone to push their limits and boundaries. Members who consistently meet with failure on the training ground will duplicate the failure on the fireground. The behavior and performance of our members will mirror our expectations for them. Messages such as "You can do it" and "I won't give up on you" communicate our belief that they will succeed (fig. 4–3).

The Four Ps to Success: Plan, Prepare, Present, and Post

Fig. 4–3. Three key messages in training are the core of what this book is about. *Courtesy:* Kim Fitzsimmons.

Finally, once objectives have been set and the participants informed of them, the objectives should *never* be changed simply because the participant is about to succeed. We have all seen this, many have done it, and more still have been the victims of it. If the participant meets the objective, even if they do so easily, it does not invalidate the exercise. It does not reflect badly on the instructor, either. Rather, it likely means that the instruction was adequate to provide the participants the skills to perform. Would we categorize a school with a high student failure ratio a good school? It is more likely that we would question the quality of the teachers. However, if you find that most members complete the exercise easily, you may need to reevaluate your needs assessment and adjust future activities. Changing the objectives during the exercise in order to ensure failure or to increase the difficulty for the participant who is succeeding is the sign of an insecure or inexperienced trainer.

Practice run of the exercise

Taking the time to do a practice run of the exercise may seem like a luxury that you cannot afford. Volunteer departments as well as career paid departments struggle to have extra time and available people. However, it is important to make every effort possible to ensure that instructor and student time is used efficiently. An instructor may choose to practice the presentation in real time, perhaps in front of a mirror. Many have found this to be a helpful way to prepare. One of the most important components of communication is body language. Seeing yourself exhibit body language that could impede learning can be very beneficial to you as a presenter.

It may be necessary to employ adjunct instructors on more complex activities. Take firefighter survival, for example. Firefighters may find themselves in situations involving entanglement, confined space egress, or being trapped unprotected above a fire. The solutions to these emergencies require distinctly different skill sets. Enlisting outside subject-matter experts is one way to address this issue. "Training up" members of your company and developing individual skill stations for them to conduct with the other participants is efficient and empowering for those members.

Finally, during the preparation stage of development, do your research. Gather supporting documentation and/or dynamic media, and vet the information. Other than the trainer being completely unprepared, nothing discredits the training message faster than using unsubstantiated data to deliver the lesson.

THE PRESENTATION

We discussed earlier the disadvantages of the no-notice drill, which is more of a test than a learning event. How, then, do you avoid creating anxiety for the trainees you wish to engage and inspire? Let members know in advance what will be covered. If you are following this four-step process, you have identified what the topic will be through a needs assessment. You have narrowed down the performance objectives based on terminal objectives sought (competence or mastery) and have confirmed delivery timelines. Now you need to get the word out. Send out an announcement to the intended audience describing the training in detail. Provide some background on the subject. If the subject was chosen due to an incident, good or bad, constructively explain the choice. Spell out the objectives. Explain the layout of the exercise and how the information will be presented. And finally, attach reference material, e.g., relevant policies, standard operating procedures or guidelines (SOPs or SOGs), and links to articles or videos that may be reviewed in advance. Poor attendance or participation, particularly for hands-on training, is often linked to barriers to training created by past failures and embarrassment. Trust is a characteristic that must be developed by the trainer, and clarity of training information and expectations is one way to gain that trust.

As the preceding illustrates, there is much work to do to make your training event a success. Some models suggest as much as four to five hours of preparation for every hour of lecture. Depending on complexity, preparing for hands-on events could take considerably more time. Once all the groundwork is done, subsequent meetings can almost be plug-and-play efforts.

A key rule for presenters is to *be punctual*. That does not mean that you arrive for the 0900 training by 0900. It means that you are set up and ready to roll at 0900. Do not make your participants wait for you. Also, only under special circumstances should you delay the beginning to allow for latecomers. This sends a signal that it is okay to be late. Such lack of accountability can spread like wildfire, affecting future training.

Sessions should begin with a review. Remind participants of the objectives and explain the layout and timing of breaks, skill station rotation, and emergency procedures. If your department requires formal safety plans for hands-on exercises, review the details of any safety concerns, mitigation factors, and how to signal that an emergency exists if one should arise.

It helps to focus the learner's attention. Review of a video, a photo of an incident, or even a recording of an incident that was the impetus for the training can serve to focus everyone's attention. This also serves to fulfill the law of readiness discussed in chapter 3. In addition, it meets one of the definitions of the law of primacy (that the task at hand is of foremost importance). Providing a detailed demonstration of the skills to be learned addresses the other aspect of primacy—the method learned first is the most readily recalled.

Another issue to be addressed during the briefing is the level of personal protective equipment (PPE) required. Exercise prudent judgment here. Members must be able to perform in the level of protection that they will use on actual calls. However, it may not be necessary to have full PPE while learning the steps, techniques, and hardware. Use caution requiring full turnouts during the learning if full turnouts are not required for actual protection. The student that is distracted by how hot he or she is in the July heat is not concentrating on the finer points of rigging a 3:1 hoisting rig. This is where progressive degrees of difficulty, discussed in part 3, come into play on the path from competency to mastery.

While we know that everyone is responsible for safety, it is still important to appoint a safety officer whose primary function is safety oversight. Ideally, this is someone with the requisite knowledge about the operation at hand to recognize unsafe acts and who has the authority to stop such acts unquestioned. That does not mean that the activity is over, only that current actions need to stop to address the issue.

Instructor to coach

Earlier in this chapter, we said that objectives need to be set, and the presentation should include instruction and demonstrations providing the information needed to be successful in the exercise. Conducting practice runs ahead of time with the adjunct instructors should have confirmed that goal. But invariably, some will struggle. That is okay. If most, however, are struggling, the needs assessment may have missed the mark determining the expected participants' baseline skills or knowledge. This may require a wholesale regrouping immediately. If only a few are having difficulty, though, the instructors must be willing to step in and coach. Most adults do not want someone to jump in and take over when they are trying to learn. It may just require a hint or reminder when students are stuck or struggling. Let them try to work it out, but do not let them work to utter frustration. Observe carefully what is going on for each participant. If need be, have the participant stop and discuss the difficulty. Help the participant talk through the steps that were reviewed or demonstrated previously. (See the Power of Self-Talk discussion in chapter 8.) This can be very constructive. Lessons learned under a certain amount of stress are very vivid. Imagine what would happen if they were allowed to fail under heavy stress. What would happen if they were permitted to leave the training site having *not* successfully utilized the skill, and in due time, they were in a situation that required that skill? The failure rate in such cases is predictable, and the penalty for failure in a real emergency tends to be severe (fig. 4–4).

Fig. 4–4. The Greek poet Archilochus said, "We don't rise to the level of our expectations, we fall to the level of our training."

Every effort must be made to ensure that the skill was completed correctly. This may mean individualized, one-on-one instruction after others have finished. Remember, we are teaching, not testing. Success is based on the human ability to train to the point that when someone is faced with a critical situation, he or she has practiced the responses so many times that it has become an immediate reaction, occurring without thought.

Rehabilitation

The days of telling members, "Suck it up, Buttercup" are thankfully behind us. Providing for rehabilitation on the fireground is becoming a part of our culture as much as self-contained breathing apparatus (SCBA) and full turnout gear. Rehab is part of most incident action plans on fire events and has been part of hazmat response for some time. It is part of *NFPA 1403: Standard on Live Fire Training Evolutions* for good reason.[1] Heat exhaustion and heat stroke are serious heat-related emergencies, and it is incumbent upon drill coordinators to provide for protection against injuries. Consideration for prudent work/rest cycles should be weighed during the preparation phase of development.

The 2015 edition of *NFPA 1584: Standard on the Rehabilitation Process for Members During Emergency Operations and Training Exercises* addresses the roles, responsibilities, and guidelines for rehab at incidents as well as at training events.[2] This standard defines the responsibilities from the incident commander (IC) down to the individual, and it provides the framework for identifying suitable rehab areas, minimum acceptable work and rest cycles, and the importance of having EMS on the scene available for help with heat-related and other emergencies.

While formal rehab is the norm during live fire training, it should not be overlooked during non-live-fire activities. Annex D of the 2002 edition of *NFPA 1403: Standard on Live Fire Training Evolutions*, states, "Tasks such as stair climbing, roof venting, and rescue operations, when performed in full gear, have an energy cost of 85 percent to 100 percent of maximum capacity and lead to near maximum heart rates."[3] It stands to reason that training for such activities will also create high cardiovascular demands, and internal heat from strenuous activities can quickly become an issue. Provide an adequate rehabilitation area that is protected from the elements where physical monitoring, rehydration, and food are available during rest cycles. Having provided that, it is imperative that you *mandate* participation in the process. It is not only prudent, as heat-related injuries and fatalities are a reality, but also it is particularly egregious when members are injured through failure to provide for rehabilitation or to follow proper actions.

POSTEXERCISE DEBRIEF

Some trainers like to use the rehabilitation area for the postexercise debriefing. This can be very effective in many cases because members are corralled in a defined area. They are almost captive. Ideally, this area is close enough to the action area that it is easily accessible but far enough away or with barriers so that subsequent scenarios by others are not a distraction. If vital sign screening is being conducted in rehab, it may be best to let that conclude before beginning the debriefing to avoid distractions.

Depending on the format for the exercise (skill stations, scenario based, concurrent operations), a roundtable discussion is usually very effective. Call on the participants to express their experiences, good and bad, before initiating your critique as an instructor. Often, if a constructive atmosphere has been established throughout the exercise, participants who are asked to share what went right or wrong or what they might have done differently are willing to be more self-critical. Conversely, going down a laundry list of performance shortcomings may make members defensive and unwilling to receive corrective advice. There is not a one-size-fits-all approach to debriefing, and group dynamics may require some adaptation.

Finally, consider publishing an after-action report, which can be very helpful moving forward as a formal record of training received, potential deficiencies identified, and corrective measures. Consider sharing the report throughout the department and with surrounding mutual or automatic aid jurisdictions. Doing so allows those who were not present to share in the lessons learned and may trigger future exercises on the subject. Obviously, the report should speak in general terms as to a specific unit's or individual's performance but should clearly identify issues and actions or recommendations for improvement.

SUMMARY

It is prudent and efficient to follow a deliberate process for development of drills or training exercises. The four steps discussed here—plan, prepare, present, and post—comprise the Four *P*s to drill-time success. Keep the terminal goal in mind from the onset. Are you introducing new information or processes? If so, the objectives must be selected to first address competency and confidence before ramping up intensity to achieve mastery. If baseline competence is already in place, more aggressive and progressively difficult objectives can be employed. Keep the laws of learning in mind when following these steps, and your success-based training model is well underway. Remember, providing the information and skills for your students to be successful is not spoon-feeding. It's called teaching. Anything less is just testing.

NOTES

1. NFPA, *NFPA 1403: Standard on Live Fire Training Evolutions* (Quincy, MA: National Fire Protection Association, 2018). This standard can be accessed from https://www.nfpa.org/codes-and-standards/all-codes-and-standards/list-of-codes-and-standards/detail?code=1403.
2. NFPA, *NFPA 1584: Standard on the Rehabilitation Process for Members During Emergency Operations and Training Exercises* (Quincy, MA: National Fire Protection Association, 2015). This standard can be accessed from https://www.nfpa.org/codes-and-standards/all-codes-and-standards/list-of-codes-and-standards/detail?code=1584.
3. NFPA, *NFPA 1403: Standard on Live Fire Training Evolutions, Annex D* (Quincy, MA: National Fire Protection Association, 2002), 28, http://www.in.gov/dhs/files/nfpa1403.pdf.

5

The Safety Plan: Not Just Another Form

WHILE IT MAY SEEM like the last thing that we need in the fire department is one more form, the formal safety plan can serve more than one purpose and should be embraced.

The fire service and the business of fighting fire is inherently dangerous. It is not possible to eliminate the risks involved. It is, however, possible to identify them, mitigate them, and minimize them. It is difficult to reconcile any injury during training, especially a debilitating or life-threatening injury that should have been anticipated and the risks minimized beforehand.

FEATURES OF A TYPICAL SAFETY PLAN

Let us examine the features of the generic safety plan provided (fig. 5–1). You may find it useful as a planning template to guide you through the four-step process outlined in the previous chapter: plan, prepare, present, and post.

The header naturally contains the date, time, and location. The first section delineates the type of exercise. The list is not intended to be all-inclusive. It may include activities or specialties in which your department does not normally participate. However, it may still be useful when planning to participate with an outside entity. It can serve as an internal document for your records and to ensure that all safety concerns are addressed for your members. If you use this example as a model to design a department-specific template, you may add or delete items or categories as appropriate. In any case, it provides a template to ensure that all concerns and particulars are addressed during event development.

Drill Objective(s)

The next section begins to form the foundation. This is where one would state the objectives. Recall that the main characteristic of appropriate objectives is that they are achievable. By that we mean that the number of objectives is reasonable for the event planned. If constructing an electronic template for internal use, this data cell should be of fixed size and format. Consider defining the cell to contain no more than three or four bullet point objectives that the participants will expect to achieve. If there will be multiple objectives, they should be related to each other. (*Note:* It is not necessary to have multiple objectives. Complexity and time may allow for only one objective, and that is fine.) Objectives should be clear, concise, measurable, and repeatable.

Description of Training

Here, you may use as much detail as you like. It need not be a script, but it should be useful as a lesson plan for a drill coordinator or adjunct instructor to follow and ensure continuity of the message. The design of an electronic version of this section should allow for this cell to automatically expand as the user types in the information. A printed version should have enough space for substantial input.

Department-Related P&P, SOP, and SOG

This section will cover departmental policy and procedures, along with standard operating procedures and guidelines. It may be helpful to design this data cell to be able to expand automatically. Depending on the complexity of the activities, there may be only a few relevant policies or guidelines, or there may be a host of them. For example, a full-scale multicompany scenario-based exercise with real-time concurrent company operations would include everything from policies governing minimum manning to apparatus placement, incident command, hose selection, fire attack strategies, and more. A review of self-contained breathing apparatus (SCBA) may be covered in a single document. In either case, list the documents that provide support for the message and the lesson. This eliminates the inevitable questions about the propriety of the skills being shared.

PPE/Equipment Required

Again, this is clearly not an all-inclusive list. Personal protective equipment (PPE) may vary depending on the terminal objective (competency or mastery) and the participants' existing skill levels. The level of PPE must be appropriate for the level of risk for injury or exposure. A requirement that full turnout gear be worn for "conditioning" to get members used to working in their gear may be creating difficulty simply for the sake of difficulty. In addition, it could create a physical hazard to participants from heat stress. If the participants are getting overheated (clinically) or are simply uncomfortable when they are trying to comprehend a complex process in your message, the PPE will create a barrier to learning.

Hazards and Control Measures

In departments that currently require the submission and approval of some type of a formal safety plan, this section may be the most dreaded among ambitious trainers. Reviewers of such plans may feel obligated to *eliminate* safety concerns on the training ground. As alluded to earlier, that may be a lofty goal. Activities that prepare you to face down the hazards of the fireground or other emergency scene eventually need to closely replicate those conditions. As such, they will present some level of hazard to the participant. Knowing this, the trainer may feel inclined to minimize or eliminate mention of certain hazards for fear that the "safety police" will shut down the training. Leaders of departments need to foster a cooperative atmosphere in this area whereby officers, instructors, and those specifically tasked with safety oversight work toward having the highest level of training possible to provide the quality of response that the public expects and deserves. The list of hazards on the sample form covers a wide array of potential safety issues. Some can be very significant, particularly if acquired sites are being utilized. Generally, such sites tend to be in a state of neglect or disrepair and may include structural issues, debris and biohazards, the existence or absence of electrical power, open drainage, and other challenges. These conditions do not automatically preclude use of the site for training. They will obviously limit the scope of use, but where hazards can be isolated and access to dangerous areas restricted, the means to do so should be listed in the space provided. Some issues may require that a sentry be posted to positively block access. The value of the training will dictate how far you may wish to go to secure an area.

Accountability

This is not an area to be taken lightly. A report published by the NFPA in 2012 analyzed firefighter deaths that occurred in training between 2001 and 2010.[1] Firsthand accounts of several of these instances revealed that the firefighters who perished were lost for a period of time, were thought to have left the building, or were not known by instructors to be in the building. Accountability systems should be used on the training ground as well as the fireground to avoid such confusion.

At the minimum, a buddy system or direct visual contact should be maintained. Passport systems or control boards utilized on the training ground help enforce their use on the emergency scene as well, and their use should be required by policy.

In Case of Emergency

The first choice in this section should be self-explanatory: *Follow your Mayday procedures!* Some departments employ a practice of some odd code to let everyone know that they have a real emergency. We know that it is a flawed concept to expect someone to think differently under extreme stress and to not respond reflexively when faced with a threat. This topic was addressed in chapter 2 regarding SNS response and in chapter 3 regarding the laws of repetition and intensity. If we accept these truths, then we know that the likelihood of calling a special code during an emergency while training and another code on the fireground is slim. The term *Mayday* should always mean a member is in immediate trouble, period! It should evoke the same response on the drill field as the emergency scene, with no exceptions.

Not doing so can result in an actual Mayday being perceived as a twist thrown into the exercise, and therefore the response may be delayed. If the exercise is to include a Mayday by design, then a variation of the signal may be used, i.e., "Mayday, Mayday, Mayday, this is a drill!" All of those who hear this, including the dispatch center, will know that no one is truly in peril.

The aforementioned NFPA LODD report notes that in some of the cases, a rapid intervention team (or crew) (RIT or RIC) had not been assigned or identified. These were specifically on live fire exercises, but it is prudent to consider this assignment if the conditions and activities planned may lead to a condition where a RIT or RIC may be deployed. Also, if resources permit, having emergency medical service (EMS) with advanced life support (ALS) capability standing by at the training site to take care of our own is a good idea. It is not sufficient to consider the EMS crew participating in the evolutions as EMS coverage. Due to their involvement in the exercise, they may not be immediately available should an emergency arise. It could be the participating EMS crew that needs to be rescued.

Communications

In the communications section of the form, you will identify the primary and secondary radio frequencies. You must also provide for review of alternate communications, such as hand signals (apparatus operation or wildland training, for example) or even rope signals for water rescue operations. Confirm, if necessary, that the frequency intended will actually be available or monitored on the day of the exercise. Ensure that all participants know the frequency assignment and that all apparatus and hand-held radios are set prior to beginning.

Resources Assigned

This section can be used to roll call participating units at the exercise as well as units held in support of the activity. Ensure that all are monitoring the appropriate radio frequency and understand what their role is should there be an emergency.

Safety Plan Notes

This section should be used for any special notes, contact information, site plan drawings, etc. Signature lines for approval, if required, ensure that the details of activities are known and are authorized.

SUMMARY

Everyone involved must understand the briefing outlining the scope of the exercise at the beginning of any training activity. Some trainers jot down briefing notes on a pad or just wing it. Others might skip the briefing all together, believing if members do not get a briefing before responding to an emergency, they should not need one on a drill. As discussed in a previous chapter, this describes a test, not a learning environment. The briefing should also include a safety briefing, which is required in *NFPA 1403: Standard on Live Fire Training Evolutions*.[2] Many of the requirements of that standard would be prudent to follow even on non-live-fire exercises.

The safety plan example presented here covers those needs. It also may serve as a planning template to be utilized during the development of a training event. Each component of the four-step plan outlined in the previous chapter is supported in this document. The objectives are identified and stated. The specific actions and events are recorded. Specific policies, operating procedures, or guidelines validate the processes and provide for the logistical needs. Logistical concerns may vary from equipment to site prep (safety), communications and emergency actions, and designation of individuals who will oversee various areas of responsibility.

As the NFPA report, *U.S. Firefighter Deaths Related to Training, 2001–2010*, concludes, among other observations:

> Since training exercises should be conducted in controlled settings, they must be designed so as not to endanger the participants. This requires that recommended safety procedures be followed. That, in combination with competent instruction, should result in the level of safety necessary to protect the lives of those participating in training.[3]

A form will not eliminate the hazards. It cannot guarantee that there will be no injuries. It can, however, provide a comprehensive approach to a vital task required of those responsible for preparing members of the fire service to protect their communities. Training must be carried out with the highest consideration to safety.

FIRE RESCUE
Drill Safety Plan

Drill Date: _____ Time: _____ Shift: _____

Drill Location: _____

Type of Training: *(check all that apply)*

- ☐ Air Rescue
- ☐ ARFF
- ☐ EMS-related
- ☐ Fire Suppression/Hose Handling
- ☐ Haz Mat/WMD
- ☐ Live Fire Training
- ☐ Search and Rescue/RIT
- ☐ Technical Rescue
- ☐ Water/Dive Rescue
- ☐ Other: _____

Drill Objective(s): *(brief explanation of objectives)*

Description of Training: *(e.g., Extricate victim from vehicle, victim search in limited visibility)*

Fig. 5–1. Example of a comprehensive safety plan

The Safety Plan: Not Just Another Form

Department Related P&P's, SOP's, SOG's: *(list number and name)*

PPE/Equipment Required: *(check all that apply)*

- ☐ Helmet
- ☐ Eye Protection
- ☐ Hearing Protection
- ☐ Gloves (Type) _____
- ☐ Bunker Coat
- ☐ Hood
- ☐ Bunker Pants
- ☐ Safety Boots
- ☐ Other: (Specify) _____

- ☐ Personal Flotation Device
- ☐ Buoyancy Compensator
- ☐ Mask/snorkel/fins
- ☐ SCBA
- ☐ SCUBA
- ☐ Other Resp. Protection (Type) _____
- ☐ HazMat CPC (Type) _____
- ☐ Radio

Hazards & Control Measures: *(check hazard AND write in control measure)*

- ☐ Atmospheric (smoke, dust, low oxygen, etc.): _____
- ☐ Combustible/Flammable Environment: _____
- ☐ Confined Space: _____
- ☐ Electrical: _____
- ☐ Elevation: _____
- ☐ Hazardous Substances (asbestos, chemicals etc.): _____
- ☐ Nighttime Conditions: _____
- ☐ Sewage/Septic: _____
- ☐ Sharp Edges / Objects: _____
- ☐ Structural: _____
- ☐ Terrain: _____
- ☐ Traffic: _____
- ☐ Water: _____
- ☐ Weather: _____
- ☐ Other: _____

DEVELOPING FIREFIGHTER RESILIENCY

Accountability: *(check all that apply)*

- [ ] Buddy System
- [ ] Visual
- [ ] Passport
- [ ] Dive Master Control Sheet
- [ ] Other: _____

In Case of Emergency: *(check all that apply)*

- [ ] Code or Signal Used: _____
- [ ] RIT Assigned: _____
- [ ] ALS Standby: _____

Communications:

- [ ] Radio/Primary Frequency: _____
- [ ] Radio/Secondary Frequency: _____
- [ ] Hand Signals
- [ ] Rope Line
- [ ] Lights
- [ ] Other: _____

Resources Assigned: *(check all that apply AND fill in designated unit)*

- [ ] Battalion/District Chief(s): _____
- [ ] Rehab Officer/Area: _____
- [ ] Rescue Unit(s): _____
- [ ] Safety Officer: _____
- [ ] Specialty Unit(s): _____
- [ ] Suppression Unit(s): _____
- [ ] Other Resources/Equipment: _____

The Safety Plan: Not Just Another Form

Safety Planning Notes: (Site Plan, Drawings, etc)

Drill Coordinator: _____
(Print Name)

Signature: _____   Date: _____

Reviewed by (Print): _____
(Battalion Chief, Training Captain, OR Field Safety Officer)

Signature: _____   Date: _____

NOTES

1. Rita Fahy, *U.S. Firefighter Deaths Related to Training, 2001–2010* (Quincy, MA: National Fire Protection Association, 2012), https://www.nfpa.org/-/media/Files/News-and-Research/Fire-statistics/Fire-service/osffftraining.ashx?la=en.
2. NFPA, *NFPA 1403: Standard on Live Fire Training Evolutions* (Quincy, MA: National Fire Protection Association, 2018). This standard can be accessed from https://www.nfpa.org/codes-and-standards/all-codes-and-standards/list-of-codes-and-standards/detail?code=1403.
3. Fahy, *U.S. Firefighter Deaths Related to Training*, 11.

PART 3: RESILIENCY

Introduction: When It All Comes Together

Dave Gillespie

EARLY IN MY CAREER, I had an incident that changed my life, and it was a catalyst for my involvement in this project. I became disoriented during a fire attack inside and apartment building that had been built in the early 1900's. My crew was up on the third floor, and visibility was a clear 20 feet with light smoke. We pulled a wall of lath and plaster to search for fire, and down came the kitchen cabinets with a loud crashing noise. The balloon frame construction was a tunnel of fire and black smoke. The kitchen went to untenable conditions. Within a split second, visibility went to zero.

I became separated from my crew while they exited, and I lost contact. Then I lost my radio. Figuring it was going from bad to worse, I reverted to my training to drop to my knees and search for the last wall. On finding a window, I signaled to the crews on the ground below to raise a ladder. As I climbed out the window, I could feel the adrenaline pulse in my neck and in my forearms as I gripped each rung.

When I got down to the street level, my Captain's angry expression and his lips moving made it look like he was talking, but I still could not hear anything. He pointed to the radio in my chest pocket, and said something like, "Don't forget to use your radio."

Although it was unknown to me at the time, the stress response of fight-flight-freeze had hit me hard. I was unfamiliar with how the body reacts to shut down hearing, dilate the pupils, cloud rational thinking in the frontal cortex of the brain, or reduce fine motor control skills like gripping a radio.

Years later, I was able to put this experience to a more constructive pattern in training to deal with stressful situations.

6

Mental Tactics of High Performance

THE CALL I had prepared for came in at 17:40, a confirmed working fire with a child trapped. Here was a call that would allow me to practice my new skills on an active fireground after 16 years as an acting captain for Peterborough Fire Services, Canada.

I recognized the address in an older neighborhood next to the General Electric plant established in 1910. The house would likely be a wartime bungalow. The neighborhood was only one mile from the station.

After the alarm tones, our crew ran to the truck. I could feel that my heart rate had jumped up to more than 120 bpm, so while putting on my bunker gear, I began box breathing. *Inhale 2-3-4. Hold 2-3-4. Exhale 2-3-4. Repeat.* Once we were on the truck rolling out of the station, I confirmed the address and route with the driver and gave assignments to my crew.

On route, the next step was to envision the neighborhood and probable bungalow. I ran a mental scenario and rehearsed the approach, seeing B-side, arriving and exiting on A-side, reading the house and layout, and then scanning the C-side. I pictured giving my crew of two rookies simple instructions to extend a 1½-inch line to the front door. The chauffeur/pump operator was experienced, so I trusted him without any verbal direction.

Before arrival I rehearsed a possible initial report in my mind, hitting on key points. There was no checklist here; it was all rehearsed. I performed one last box breathing cycle: *In 2-3-4. Hold 2-3-4. Out 2-3-4. Hold 2-3-4.*

We pulled up to a bungalow, approached from B-side, exited A-side, and read the window layout. I then gave my initial report as I had visualized just seconds earlier. I was approaching the C-side before making our entry when the mother came around the house screaming, "My son, my son! He's inside!"

The adrenaline was pumping, and we were rocking to get to the front door. The two rookies had stretched the attack line, and we met on the front step. As they approached the front door, they were fumbling with their regulators.

I directed them, "Slow down. Just breathe." This helped me as much as it helped them.

Inside it was hot and black, with visibility at six inches. Training had prepared me for this. A few months earlier, we had conducted dozens of training search scenarios in the dark in the new training prop, with loud sounds, essentially doing stress inoculation drills. On our way into the house, my sit-rep (situation report) included "P1 on fire attack and search, making entry on A-side, searching using right hand rule, PAR 3."

As we looked for the child, I kept repeating to myself, "We've got this." Positive self-talk is important, and we knew the odds were against us, but we went in anyway.

The incident command's tactical objective was to do a primary search of the house. But to do that, we needed to search one room at a time for a little boy. Clearing each room evolved into a micro goal-setting exercise. By handling one room at a time, we segmented our search into two-minute chunks, which allowed us to stay focused on looking for a small boy.

At each room, I paused from crawling on my hands and knees, sat up, and rolled back on my feet into a kneeling position. Feeling the weight of the SCBA on my shoulders, I stayed under the thermal layer. This allowed me to expand my lungs and box breathe more easily. *In 2-3-4. Hold 2-3-4. Out 2-3-4. Hold 2-3-4.*

Using two methods of breathing significantly slowed down my respirations and heart rate. I used the Reilly breathing technique of humming for two cycles and then box breathing during the pause as my crew was getting repositioned in the hallway. My heart rate fell into a calm range immediately, allowing me to talk calmly to my crew, think of our next move, and communicate clearly to command by radio.

We cleared the main floor and provided a situation report at the top of the stairs leading to the basement. That's when I saw the woman's son sprawled at the bottom the stairs. To our surprise, he appeared to be about an 18-year-old, 200-pound "little boy."

Who would think that a technical firefighter would be using mental rehearsal, self-talk, and tactical box breathing techniques? I never expected it would be when leading a first-due engine company.

Again, I practiced one cycle of box breathing as we went down the stairs. One of my rookies assessed the young man and called out, "No vitals." I radioed, "Command, this is Portable 1. We have found the victim at the base of the stairs on C-side. We will be transporting him to the front door. Have EMS ready." Command acknowledged.

As we lifted the young man into a two-man carry, we stayed energized but methodical. We carefully extracted him up the stairs where we met my other rookie. He had operated the nozzle to extinguish the kitchen fire. He was now leading us out to the front door on A-side, where we passed off the patient. EMS began compressions right away on the front lawn. They regained a pulse quickly and transported him to the local hospital.

Our crew was ecstatic with the rescue and a knockdown in 15 minutes. The biggest compliment came when the IC came over and commented, "Wow! You were calm on the radio. I heard everything you said on interior operations and could easily follow you. It was like you were on a golf course—calm and controlled."

Using the Big Four

As seen in sports and the military, these tactics have changed the way I now manage myself for high performance:

- A quick mental rehearsal
- Constructive self-talk
- Micro goal-setting
- Tactical breathing (see chapter 7)

These are my tools for every call.

TAKING CONTROL OF YOURSELF

The average age of a US Olympic athlete is 26.[1] The average age of a US Special Forces (SF) operator is 30. For elite SF operators, average age can climb to 36 to 40.[2]

From a fire service perspective, why does it take 15–20 years for a high-performing, tactical, athletic firefighter to be considered experienced? Why does it take until age 40 or 50 for firefighters to finally arrive at long-term mental well-being, long after the point at which they have mastered their technical skills and physical actions and have learned to manage physiological reactions to high-stress situations?

Question: Why does it take so long for a firefighter to be considered experienced?
Answer: We do not train in resiliency. Others do. We do not.

Unfortunately, fire service training programs often focus on hard skills and react to the effects of long-term stress only when the mental armor starts to crack. Relying on employee assistance plans to support us after the fact is simply waiting for a firefighter's natural resiliency to fail.

What if we built stronger resiliency?

RESILIENCY

Resilience is defined as "the ability to persist in the face of challenges, and to bounce back from adversity."[3] No one escapes a career as a firefighter without absorbing tragedy. Allowing tragedy to be your teacher rather than your master has a direct impact on your mental health.

SITUATIONAL AWARENESS

Situational awareness is a key component of resiliency since it is an internal and external tool alerting us to circumstances, dictating how we approach a task. For example, knowledge of strategy and tactics combined with knowledge of building construction and fire behavior help to develop our external awareness on fireground situations. This may greatly benefit a quick awareness of conditions, rather than slow awareness or none at all.

Experience and education play a key role when a firefighter must determine whether a situation is manageable or unmanageable. Recognizing signs of the potential flashover can allow early mitigation by extinguishment and/or ventilation. There are many more situations such as search and rescue (SAR), vent, enter, search (VES), and forcible entry that require the same attention to detail that improves your awareness levels.

Internal situational awareness is as important as external situational awareness. Firefighters must maintain situational awareness of their temperature. Heat stress may contribute to heat exhaustion, heat stroke, and/or heart attack.[4] The length and intensity of exertion can be directly tied to heat stress and the associated complications. Breathing has proven to be a valuable tool when it comes to regaining control of internal situational awareness.[5]

A Mayday can have a psychological and physiological effect on a firefighter. Actions by the sympathetic nervous system will increase blood pressure, heart rate, and respiratory rate, while decreasing cognitive function. Breathing has shown to decrease blood pressure, heart rate, and respiratory rate, as well as improve cognitive function.[6] Knowing what is going on inside you and around you gives you the power to start managing it.

When new recruits are welcomed to the training academy, they are introduced to technical task-level skills in their first few months. Next, they learn how the pieces fit together on a job. Some stations are constantly busy with rolling calls, while other stations are so quiet, they seem like a retirement home. It is these quieter stations where station officers are especially tasked with keeping firefighters' skills sharp. Training could potentially be focused on simple hard skills like truck equipment, saw maintenance, and ladder use. Hard skills are always the preferred training.

On the job, management of mental health and mental resiliency is left up to each individual to learn at his or her own pace, often from a senior firefighter. However, this situation has too many variables and wastes valuable time. Only a few progressive departments in the United States and Canada operate a consistent mental health program to prepare their people. So, it is our job to provide those lessons here.

MASTERING THE SELF

If the human resources department or fire department recruiters have done their job properly, they hire a rookie with a positive attitude and relatively good people skills. The new probationary firefighters are expected to maintain these basic skills throughout their career. Their colleagues and the public expect them to eventually master technical skills in lifesaving techniques, such as first aid, search and rescue, fire suppression, and auto extrication. This process enables the chief to say the team has the right technical skills to put out fires, rescue someone, or save a house or forest with competence and high performance in stressful situations.

The psychological skills that are expected to match those high-performance levels are rarely taught in the fire service. The psychological practices of elite performers have become mainstream training in other demanding professions, namely in law enforcement, US Special Forces, professional athletics, and even amateur athletics, such as the Olympics teams. This is not true in the fire service, however, and that must change. The fire service is in the ideal position to incorporate principles of high-performance psychological training.

THE THREE LEVELS OF SELF-AWARENESS

The three levels of self-awareness for individuals who perform specialized tasks are basic, intermediate, and high performance. A simple paradigm for learning levels and skill sets appropriate for managing one's own mental and emotional performance highlights how we can accelerate learning as opposed to merely relying on years on the job (fig. 6–1).

Fig. 6–1. Psychological self-management paradigm
Courtesy: Dr. Jack Lesyk, Ohio Center for Sports Psychology.

Level I: Basic psychological self-management skills

Recruit academies function with probationary firefighters at level I, where they promote goal setting by completing job performance requirements (JPRs), with quizzes and exams each week.

When starting their careers, most average firefighters operate at level I by adopting basic characteristics to help them fit into the team. Most firefighters will stay at level I, much like a rookie with experience, proud of eventually becoming a "grunt."

They will stay at level I unless they are exposed to competitive sports, are dedicated to learning a complex physical skill (music or a hobby), or participate in a cognitive counseling program or formal training sessions on resiliency. Characteristics of rookies include maintaining positive attitudes and motivating themselves and others by supporting team members. Other activities and attributes include cleaning the station, hitting the gym, being diligent with training on fireground basics, and staying focused at a fire scene.

One of the central pillars of the fire service is life-long learning. To paraphrase the late Phoenix Fire Department Chief Al Brunacini, we seem to be "going to school forever."[7] When firefighters follow this ethic of constant improvement, they actively pursue learning new skills or promotions to shift instructor or officer rank. In doing so, they start on the path to developing cognitive and emotional skills appropriate to a high-performance and seasoned firefighter, fire medic, or rescue technician. These steps usually result in more mental training, preplanning scenarios, and mental rehearsal of potential incidents. All of these actions will prepare them to function at a higher level of self-management skills.

Level II: Intermediate skills

Firefighters must be honest concerning their own strengths and limitations. They must identify gaps in internal "software" skills and seek out opportunities to grow in their mental techniques to engage a higher skill level, such as level II. Firefighters who want to practice basic sports psychology must begin experimenting with constructive self-talk, which requires personal reflection, choosing key phrases, and practicing self-talk with purpose. At level II, a firefighter also manages his or her skills in mental imagery or mental rehearsal to master higher levels of mental toughness.

Choosing a visual point of view (POV) for mental rehearsals is one consideration, such as whether the POV is from a first-person, bird's-eye view, or third-person perspective. The next step is to take your POV and develop a mental plan of what desired action you want to see yourself perform competently. It might be a first-person view of yourself doing CPR or a bird's eye view of yourself using auto extrication tools, as seen from above the motor vehicle crash. It could also be a third-person view of yourself watching the crew's actions at the fire scene, as seen from the barrier tape.

When a high-performance athlete reaches level II, they are ready for the national level of competition. When firefighters, police officers, or special forces operators reach this level, they are usually recognized for their competency by becoming the go-to person, an instructor, or possibly even the company officer.

Level III: High-performance skills

Components of level III focus on managing anxiety, managing emotions, and improving concentration. It is common to recognize top Olympic athletes possessing these abilities, but it is not common to see how they get there.

Olympic coaches now employ the sports psychology techniques of arousal control, breathing, and anticipating change in dynamic environments as part of a daily routine for their athletes, often with multiple short sessions each day. For example, Michal Staniszewski, the coach for the Canadian Olympic whitewater kayaking team, tasked his team of athletes with three-minute sessions at key times of the day: when waking up, while having a morning shower, while sitting alone, and while quietly preparing to sleep.

Fig. 6–2. A whitewater kayaker practices mental rehearsal skills at level III of high-performance skills.

For these whitewater athletes, the arousal control sessions may apply to their sport of whitewater kayaking, gym training, crowd/noise distractions, and emotional control to external events (fig. 6–2). Like many Olympic-caliber athletes, they are dedicated to practicing this skill every day in order to master it. Many athletes report using these techniques beyond sports applications in their professional work and personal lives.

Beyond arousal control and breathing, level III performers can manage anxiety by identifying what effect a given situation is having on the body. For example, "The tightness of the self-contained breathing apparatus (SCBA) facemask is affecting my breathing."

By identifying or labeling the anxiety causing the situation, the person can use reasoning skills to solve the problem faster. Here is an example of self-talk and effect-labeling: "The tightness of the SCBA facemask is making it hard to breathe. Hmm…No—since the SCBA is positive pressure—my nervousness is what's making it hard to breathe. By labeling this, now I can see I am okay. Slow down. Breathe easier."

HOW SURGEONS PRACTICE SKILLS TO PREPARE

An example of how people develop their mental rehearsal and self-talk was illustrated with a 2011 study in which researchers attempted to improve how novice surgeons learned and acquired new surgical skills. Their study, "Mental Practice Enhances Surgical Technical Skills," compared the skills of 20 novice medical surgery students who practiced surgery through mental rehearsal against a control group of 20 similar students doing identical tasks live in a lab.[8]

The skill was focused on laparoscopic surgery using a thin lighted tube through a small incision in the human belly to look for cysts or to remove gallstones or the gall bladder. The study found that students who performed 30 minutes of mental rehearsal prior to a virtual reality surgery were far superior to those who simply studied and performed the same surgery in the lab. They were able to consistently outperform their counterparts in five different surgical exams by the end of the two-month experiment.

The improved surgery skills correlated with better quality of performance and increased confidence to learn more advanced surgeries. (Specific strategies are discussed later in this chapter.)

HOW SPECIAL FORCES PRACTICE SKILLS TO PERFORM

US Navy Commander Eric Potterat of the Special Warfare Development Group (DEVGRU), SEAL Team 6, once interviewed an Olympic coach who said, "The only difference between a gold medal, and no medal, is the six inches of grey matter between their human ears."[9]

In 2008, Dr. Potterat and the psychology division of the US Navy Development Group (DEVGRU), which handles SEAL team training and operations, began to transfer the principles of sports psychology into their psychological training model for Basic Underwater Demolition/SEAL (BUD/S school). They were experiencing an attrition rate of 80%, which meant only 20 out of 100 candidates would pass to reach BUD/S graduation. Regardless of their physical ability, whether they had been athletes of water polo, rugby, triathlons, or college football, it was not the hardware of their bodies that was contributing to their failure. It was the three pounds of grey matter that made up their software.

By 2009, they were implementing the principles of sport psychology and the four major psychological principles, known as the *Big Four*, to develop their mental toughness. It was one tool that helped increase their pass rate to one-third (fig. 6–3).[10] While this may seem a small gain, it demonstrates that with well-practiced skills, those who might otherwise fail have a chance to influence their software, or thinking, to become successful.

Fig. 6–3. Navy Seal candidates are introduced to the Big Four: mental rehearsal, self-talk, goal setting, and arousal control (breathing) when performing high-risk maneuvers.
Courtesy: US Navy.

The Big Four principles are based on sports performance psychology ideas that have existed for more than 100 years. Only since 1970s, however, have they been formally recognized in high-performance sports and cognitive behavior therapy (CBT). CBT is a field of personal psychology, developed by Dr. David Meichenbaum (University of Waterloo, Canada), which is now studied worldwide in behavioral psychology.[11] Dr. Meichenbaum refined and developed standard practices that, when applied, can positively alter a person's ability to overcome mental obstacles and enhance performance.

As mentioned at the start of this chapter, the Big Four are mental rehearsal, self-talk, goal setting, and arousal control (management of visual input, hearing, and breathing). Even as we maintain control of our internal mental state, sometimes our external environment distracts us or causes a change in heart rate and breathing.

Visual input can be addressed by limiting what you choose to see, such as changing the direction of your view or closing your eyes. An example of this might be a company officer arriving at a scene involving a traumatic death and choosing to shelter a rookie from the sight.

Control of your hearing to reduce arousal might include putting on noise cancelling headphones to limit stimulation, such as might be done by a pump operator or an incident command aide. This may also involve donning ear plugs or leaving the area of the noise stimulation, if possible.

The use of mindfulness training noted in later chapters can also play a role in managing arousal state while remaining in the environment.

When using mental rehearsal, a person can prepare and see themselves during a time when they will be in a stressful situation. This allows the brain to see beyond any threat and to create workable solutions, enhancing self-confidence, motivation, and a belief in one's own ability to cope. It also stimulates the generation of neurons in the brain (neuroplasticity).

Much like an athlete engages in interval training to build their physical ability, a person can change with small doses of stress to help the brain adapt. They accept opportunities or situations that cause low levels of time-limited stress. They use them as a foundation to progressively build to more complex exposure to stress, and it becomes stress inoculation training.

When special forces operators (US Navy SEALs, Delta Forces, Rangers, or Air Force Pararescuemen [also called *PJs*]) train, they are exposed early in their special operations forces (SOF) training to formal multisession courses on mental rehearsal, self-talk, goal setting, and arousal control. This sets the stage for training experiences with progressive stress inoculation, such as ocean swimming, winter travel in −40°F weather, and high altitude–low opening (HALO) skydiving. These experiences teach initial mental coping skills integrated with physical challenges that harden the candidates' resiliency so they can face greater adversity, not just to survive, but to thrive.

When special forces operators can function at level III, they are called *operators*. This is both a sign of accomplishment of technical competencies and respect for their ability to persevere. The fire service has similar people who would qualify for a special title due to their technical proficiency and ability to persevere and be resilient. This "operator" capacity is usually someone who has 15–20 years or more of fireground experience and is known as the go-to person, or solid brother or sister who outperforms others.

Resiliency is the ability to adapt to stress and achieve successful outcomes in the face of extremely challenging circumstances. Special operations forces (SOF) and law enforcement recognize the power of resiliency to enhance performance and mitigate mental illness by building stronger people. It is time for the fire service to go beyond relying on the inherent skills of the new recruit or seasoned professional. It is time to move resiliency training from the sports field and military battleground to the fireground. If firefighters took their training as seriously as the elite Navy SEALs, wouldn't we be mentally and physically better prepared for the worst fire of our lives? Would you become more confident when responding to a fire that you had been practicing on for months?

We can accelerate our learning and competence, such as technical rescue training, using the same cognitive skills as special forces operators and police specialty teams (fig. 6–4).

Fig. 6–4. A firefighter practices mental rehearsal by knot tying. *Courtesy:* Dave Gillespie.

With a wide variety of likely traumatic calls, no career or volunteer firefighter manages challenging circumstances without absorbing tragedy and seeking ways to move forward. Challenges create opportunities for tragedies to be your teacher rather than your master, and they directly impact your mental health over a long career.

Definitive research findings from high-performance occupations such as law enforcement and military combat illustrate how perception of risk, increased heart rate, and increased respiration can lead to stimulation of the SNS response. The change in bodily functions related to the stress responses of fight, flight, and freeze impair cognitive function. Stress can reduce fine motor control and inhibit hearing, perception, and verbal communication abilities, not to mention cause sudden urination and defecation due to extreme and sudden stress.

With traditional fire training, most firefighters expect their ability to handle stress is either provided through prior life experience, initial recruit training, or on-the-job training. It can be simply an aggregate of stress-management techniques learned from years on the job. Fire instructors and training officers have a history of training to the initial stages of stress-induced responses, where the firefighter completes only the mildest training exercises or easily achieved simulations on the training ground.

It is rare that physiology is pushed beyond a comfortable level, and when it is, the firefighter tends to lose competency and revert to a base level of training.

Line-of-duty deaths and near misses are included in the hundreds of firefighter deaths and mental injuries resulting from people being exposed to elevated risk and high-stress situations, as documented by NIOSH and discussed in chapter 2. Victims become lost and perform unusual actions that lead to negative consequences or loss of life. Dozens of LODD reports have concluded that these situations resulted in a loss of situational awareness, cardiac-induced anxiety, or unexplained behavior.

The difference between sports and the fire service is that in the Olympics, an athlete might lose a competition. In the fire service, a firefighter might lose a life or become severely traumatized for years after the incident.

We can change from failure-based to success-based training models and build more resilient firefighters. The change will require proper information, guidance in training on mental awareness, tactical breathing control, and mental tactics to build operational resiliency.

NOTES

1. Alyson Hurt, Bill Chappell, and Brittany Mayes, "What Team USA Looks Like: A By-the-Numbers Look at America's Olympians," *The Torch*, National Public Radio (August 11, 2016), https://www.npr.org/sections/thetorch/2016/08/11/487838010/what-team-usa-looks-like-a-by-the-numbers-look-at-america-s-olympic-athletes.
2. Rob Verger, "Newsweek Rewind: Inside SEAL Team 6," *Newsweek.com* (May 1, 2014), http://www.newsweek.com/newsweek-rewind-inside-seal-team-6-249319.
3. Karen J. Reivich, Martin E. P. Seligman, and Sharon McBride, "Master Resilience Training in the U.S. Army," *American Psychologist* 66, no. 1 (January 2011): 25–34.
4. Denise Smith, Jacob DeBlois, Jeannie Haller, Wesley Lefferts, and Patricia Fehling, *Effect of Heat Stress and Dehydration on Cardiovascular Function* (Chicago: First Responder Health and Safety Laboratory, Health and Exercise Sciences, Skidmore College, 2015), https://www.skidmore.edu/responder/documents/smith-dhsS10-fs-report.pdf.
5. Ibid.
6. Ibid.
7. Alan Brunacini, "Going to School Forever," *Fire Engineering* 169, no. 7 (July 1, 2016), https://www.fireengineering.com/articles/print/volume-169/issue-7/departments/bruno-_unplugged_/going-to-school-forever.html.
8. Sonal Arora, Pramudith Sirimanna, Rajesh Aggarwal, and Ara Darzi, "Mental Practice Enhances Surgical Technical Skills: A Randomized Controlled Study," *Annals of Surgery* 253, no. 2 (February 2011): 265–70.
9. Eric Potterat, interview by John Marx, "Train Like a U.S. Navy Seal," *CopsAlive.com* podcast (May 9, 2013).
10. Rorke Denver, *Damn Few: Making the Modern SEAL Warrior* (New York: Hyperion Publishing, 2013).
11. "Cognitive Behavioral Therapy," *Wikipedia* (2018), https://en.wikipedia.org/wiki/Cognitive_behavioral_therapy; "Pioneer of Cognitive Behavioral Therapy," *Waterloo Stories* (July 5, 2012), https://uwaterloo.ca/stories/pioneer-cognitive-behavioral-therapy.

7

Building Mental Toughness

SPECIAL OPERATION FORCES INSTRUCTORS and SWAT team trainers have advanced their training programs in recent years to include elements of resiliency. These trainers specialize in crisis situations and emergencies and have incorporated techniques from other fields, such as college and professional athletics, to improve performance in extreme stress situations. A few examples include the following:

1. Bob Bowman, US Olympic swimming coach, employed stress inoculation activities to prepare Michael Phelps to manage stress in unexpected situations. These exercises encouraged Phelps to develop contingency plans, and he ultimately became the Olympic medal record holder, with 28 total medals.[1]

2. Steve Nash, a point Guard for the Phoenix Suns NBA team, practiced three mental rehearsals before his real freethrows, bouncing an imaginary basketball and moving his hands through the air. This strategy was one factor in his success, and he became the leading National Basketball Association (NBA) career record holder in free throws, making 90.4% of 3,400 regular season free throws.[2]

3. The Seattle Seahawks, along with their head coach, Pete Carroll, became one of first National Football League (NFL) teams to openly discuss their utilization of mental training, stress inoculation, and mindfulness to prepare for the 2013 Super Bowl. Now these strategies are employed by other teams, coaches, and athletes.[3]

Yet, the fire service has no established practices for building mental toughness. For purposes of this discussion, these tactics and more advanced techniques are divided into two categories: mental and physiological tactics.

Since the 1980s, mental training has developed from an idea to a standard practice in high-performance athletics. As discussed in chapter 6, the focus has been on four primary skills that constitute the core of all successful non-technical mental training programs:

- Mental rehearsal
- Self-talk
- Goal-setting
- Arousal control

Spectators watched in awe as Phelps won 14 gold medals in a single Olympics. Most people attribute that success to his technical training in the pool. One aspect Phelps openly credits, however, is his 30-minute mental rehearsal before each race. It provided him opportunity to visualize the process of walking on the deck, listening to his music track, doffing his training uniform, approaching the block, stepping up and onto the block, and hearing the crowd, while maintaining his intense focus.

A firefighter may have a similar experience when mentally preparing for fire attack. The firefighter might mentally review hearing the sound of the siren, smelling the fire, feeling the weight of the hose and the weight from the SCBA, and so on.

Bowman's simulation training of swimming in a darkened practice pool helped Phelps maintain his composure when his goggles began leaking as he swam the 200-meter butterfly race at the Beijing Olympics. Despite not knowing where his opponents were, or his lead or catch up time, he counted the strokes and counted the laps. Most would panic in a similar situation. But Phelps knew when to make his final push to overcome his competitors. When he reached the wall, he ripped his goggles off to see WR (World Record). He took the gold. After the race, a reporter asked what it had felt like to swim blind. Phelps replied, "Like I imagined it would."[4]

Firefighters can employ Coach Bowman's practices to anticipate problems and build contingency plans in various ways in the fire service. The practices could be employed in many different training situations, such as the following:

- When responding to a call, they discover the route is blocked by road construction and immediately plan a contingency route to the scene.
- On arrival at the scene, prior to entering the door, a firefighter test flows the nozzle, finds something is blocking the water pattern, and then calmly works through the scenario to a successful conclusion.

- While working through a primary search, a firefighter's face mask fogs up and visibility won't clear. The firefighter continues working the room, following the team's search plan and contingency plans.

Unlike sports, fire rescue work is mission specific. It has the potential to result in saving lives and protecting team members from debilitating injuries or line-of-duty deaths. Being mentally prepared with contingency plans and maintaining composure is critical to being effective on the fireground where people are depending on you.

TACTIC #1: MENTAL REHEARSAL

The concept of *mental rehearsal*, or positive visualization, can be defined as the conscious creation of mental or sensory images to enhance your goal, whether work, play, or everyday life. It is the deliberate attempt to select positive mental images or a sequence of moving action images to influence how your body responds to a given situation. Firefighters will be asked to work in reduced or zero-visibility and chaotic environments. We can apply similar methods of mental rehearsal and practice them effectively when there is downtime.

Positive visualization is not new. For years people have watched athletes use visualization exercises to master their routines in sports. Much like a football player going through the motions on the sidelines, or a gymnast walking through a routine off the mat, a firefighter can also do rehearsals. Rehearsals are valuable whether training as a team or individually in the apparatus bay.

Today, it is not unusual to see an athlete standing with eyes closed, arms outstretched, and body rotating to repeat a chosen sequence of physical actions to prepare for success.

In 2014, a group of researchers performed a study of 18 SWAT officers as they performed a five-day scenario training program. Each was outfitted with a Zephyr BioHarness to measure physiological reactions (figs. 7–1 and 7–2).[5] While listening to audio recordings of critical incident scenarios (car chase, armed robbery, violent mental health person, domestic violence, and murder scene), the officers practiced new resiliency techniques. These techniques consisted primarily of controlled breathing and positive visualization. The American, Canadian, and Finnish research team found that "SWAT members were able to significantly reduce their average heart rate and improve their ability to engage in controlled respirations, even as scenarios become more graphic."[6]

DEVELOPING FIREFIGHTER RESILIENCY

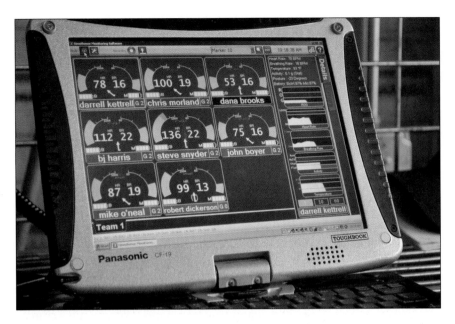

Fig. 7–1. Researchers are using biometric devices such as the Zephyr BioHarness to measure heart rate, breathing rate, body temperature, and workload. The fire service will see more of this in the future.
Courtesy: Medtronic.

Fig. 7–2. In a low-tech method, a firefighter can replicate this training using a standard Fitbit and smartphone app. A commercially available sports heart-rate monitor can also track firefighters' sleep patterns.
Courtesy: Dave Gillespie.

Firefighters can replicate this resiliency training by using carefully selected audio recordings from various fire scenes. These can be drawn from local agency recordings, YouTube, or specialty fire-based 911 radio websites. It is possible to train firefighters in the classroom or through practice simulations, where there is opportunity to introduce tactical breathing skills. Then positive visualization skills can be integrated in small steps. Using a heart rate monitor or other means of measuring changes in our vital signs can help us manage our stress response.

TYPES OF MENTAL REHEARSAL

Many people in the fire service are unaware they practice mental rehearsal. Drawing from sports psychology, Dr. Donald Meichenbaum defined *mental practice* as "the rehearsal of a motor skill in the absence of any gross muscular movements."[7] The concept is practiced in various loose forms in the fire service. It may be mentally practiced by operating a pump panel with eyes closed, walking around the engine to identify all the equipment inside the cabinets, envisioning forcing a door, or verbally rehearsing fireground size-up scenarios and initial radio reports (fig. 7–3).

Fig. 7–3. A fire officer uses mental rehearsal as a tool to practice the initial arrival report, which prepares officers to issue command decisions in a calm, coherent manner. *Courtesy:* Dave Gillespie.

Training academies do not typically train recruits to use this powerful tool, and instructors who are not trained in delivering mental rehearsal tend to focus more on a skill set of specific techniques. It is more common for instructors to teach students how to use fire tools, develop techniques, and drive apparatus than it is to have an instructor coach students on how to develop mental rehearsal techniques as a supplement to their hands-on training.

Mental rehearsal can be a beneficial practice for firefighters. Firefighters are action-oriented people who work in varying environments and deal with dynamic situations. The men and women in the fire service are progressive and adaptive people. When firefighters are shown the benefits of mental rehearsal techniques, they can quickly adopt game-changing strategies to be more effective on the fireground. Mental rehearsal can be a powerful tool for the firefighter.

As US Army Ranger Lieutenant Colonel David Grossman points out in his book, *On Combat*, the mental rehearsal of drawing a gun out of its holster and calling out, "Police...Don't Move!" must be rehearsed thousands of times.[8] This speeds up the response sequence whether the officer is standing, sitting, prone on the ground, or in the car. The effectiveness is seen in how quickly the officer can react to a threat, relying on his or her automatic trained response. Progressive law enforcement agencies have taken this further for their elite teams, like the FBI's Hostage Rescue Unit, which trains in Quantico, Virginia. Federal law enforcement training centers in Orlando and Los Angeles also prepare their teams by using different versions of imagery or mental rehearsal training for their elite teams to complement their hands-on scenarios and practical training.[9]

The terms *mental imagery, visualization,* and *mental rehearsal* may sound similar, but they are very different. Once you understand the difference, you can choose the tool most appropriate for the skill set you want to master.

Mental imagery

Mental imagery is the depiction of stationary situations, like a photograph. It could be a still image of the locking system or hinges of a door, roof types (hipped, gable, gambrel), or even a pump panel. The advantage of a still image is that it allows you to constantly review the components of the image, similar to a slide in a PowerPoint presentation. In the case of the pump panel, you may mentally locate items such as the throttle, compound gauge, and various valves and pressure settings without looking at the panel.

Visualization

Visualization is the process of imagining action-orientated tasks with an outcome, whether positive or negative, successful or unsuccessful. Proper visualization should be done in a natural sequence of events that is positive. While learning a task, we often allow the learner to fail. If that is not immediately followed up with proper execution, however, the negative mental visualization of the failure becomes the default visualization. By intervening with the correct action, the firefighter's default experience becomes positive. This is key for visualization, too. If a firefighter visualizes a series of actions out of sequence, or if the visualization becomes negative, he or she must quickly regain control and replace it with the correct visualization, which then becomes the default experience (fig. 7–4).

Fig. 7–4. Two firefighters practice their quick tool placement, visualizing the correct position to forcing a door.
Courtesy: Peterborough Fire Services.

Mental rehearsal

Mental rehearsal is the primary successful tool to engage the visual skills to aid in perfecting a performance. It is a sequence of specific physical actions that is constructive and is assembled in a sequence that will lead to a preferred outcome. It is not a random visualization but a determined rehearsal to develop excellent response skills.

These rehearsals are only limited by your imagination as to how and where to use them, such as setting up the incident command system on a command board, testing nozzle pattern before entering a structure fire, and rehearsing the quick set up of the extrication power unit and cutters for a simulated car accident. Mental rehearsal practice is beneficial when performing a primary assessment, preparing drug intervention, reviewing how to approach a drowning victim in the water, or responding to a downed firefighter on a Mayday call.

It is common to view your mental imagery in color and in action. Dr. Gary Klein, known for his recognition-primed decision model (RPDM), speaks about building a photo album or slideshow of images to rely upon when faced with challenging situations. He found that activating your memory banks is more effective when you contrast colors and make vivid mental images. A detailed recollection could be from an accident scene or a water rescue. An accident will offer many opportunities to use color to make images vivid in your visualization, while a water rescue may be more challenging. The idea is to utilize all the senses and to include visual cues such as color to enhance your visualization practice. This may include the clear visualization of the vehicles involved, such as the big yellow bus and the small red pickup truck. For a recollection of a water rescue, it may include the red lifejackets, the yellow rope, and the color of the boat. Also, recall the sound of the water and the smell of the boat's engine.

Seeing color enables you to associate the color with the object, such as a silver Halligan, a yellow axe handle, or the green light in your face mask's heads-up display. Each color reference should be significant to your visualization (fig. 7–5). The color of the axe handle may indicate that it is the one used to break open a door for search operations. The green color in your heads-up display signifies that you can continue to work because you have ample air in your cylinder, or it may allow you to evaluate how much further you can go into a structure. These are examples of how to make your visual training more realistic and vivid.

Building Mental Toughness

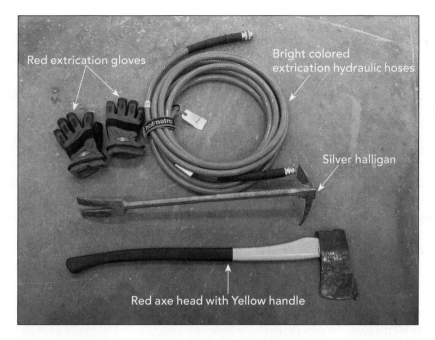

Fig. 7–5. Examples of colored fire equipment to reinforce vivid memories and mental rehearsals.
Courtesy: Dave Gillespie.

THREE POINTS OF VIEW

When establishing your mental imagery scenario, first choose the view point from which you will see the action (point of view or POV). View yourself in action as if you are directing a movie, placing the camera at the POV that grants the best advantage for training and seeing the situation unfold.

First-person point of view

Choosing to see yourself in action can take different forms, with many people performing external visual imagery as if they are viewing events live from their own eyes. This is called *first-person POV* (fig. 7–6).

In a 2013 study from the Institute of Psychology of Elite Performance in the United Kingdom, researchers found there was significant improvement in accuracy and task effectiveness by using internal visual imagery. This includes viewing one's actions from one's own upright or seated position (first-person POV), or from slightly behind one's self.[10] The use of technology such as body cameras and helmet cameras may explain some of the recent popularity for seeing action this way.

DEVELOPING FIREFIGHTER RESILIENCY

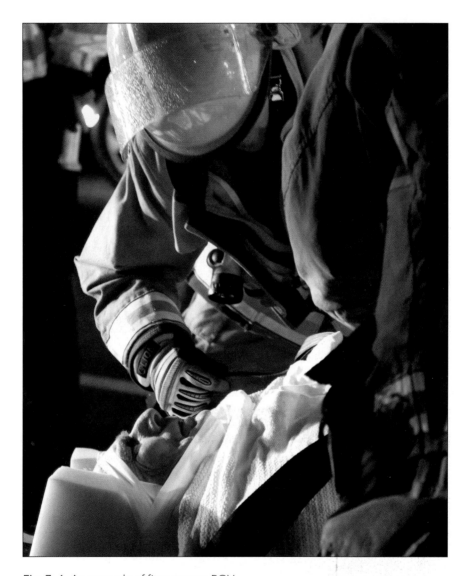

Fig. 7–6. An example of first-person POV
Courtesy: Clifford Skarstedt, *Peterborough Examiner.*

Drone point of view

Some athletes envision a flying drone or aerial view because it allows a mental rehearsal to see what everyone is doing at once, and they can zoom in on their role. This POV often can be developed and utilized for large-scale situations with many moving parts or a large footprint, such as a motor vehicle crash (MVC), plane crash, or larger structure fire (fig. 7–7). The use of drones in the fire service should facilitate adopting this POV, as it can be viewed as seeing things as if you are controlling a drone. The firefighter can shift his or her aerial mental picture left or right and zoom in or out to achieve the desired viewpoint.

Fig. 7–7. Example of bird's-eye-view drone POV
Photo credit: courtesy of Robert Horton, Borealis Productions, Canada.

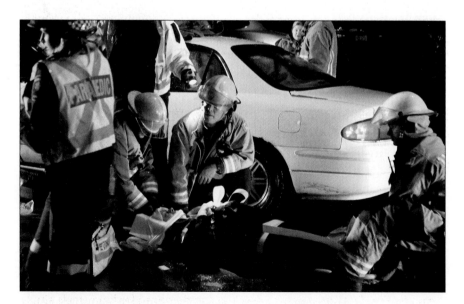

Fig. 7–8. Example of third-person POV
Courtesy: Clifford Skarstedt.

Third-person point of view

Some athletes studied preferred to view their interaction on the competition ground from a third-person POV, such as watching from their coach's position or that of a spectator or television crew recording their actions from a distance. The choice of imagery perspective will be up to each individual based on personal preference and the circumstance in which the POV is being applied (fig. 7–8).

DEVELOPING A REALISTIC GOAL FOR A MENTAL REHEARSAL PLAN

The first step is to choose the circumstance that you wish to develop a plan for, such as a challenging situation in part of your district that you respond to frequently. Allow the situation to evolve in your mind to make it easier to recreate the environment. Perhaps there are townhouses or low-rise apartments where your crew has frequently responded to fires or there is an intersection with regular MVCs.

Determining what evolution you will perform allows you to put the actions into proper sequence. Establishing a plan requires answering three questions:

- Is this a skill performed by one person who directs the situation?
- Is this a skill performed by two or three people, with each having a role?
- Is this a comprehensive skill that requires detailed and frequent updates to the actions?

Once you choose a specific skill or evolution to master, keep it simple. Examples of skills where mental rehearsal would be helpful include learning the procedure for calling a Mayday or activating a personal alert safety system (PASS) alarm, using a radio, advancing a hose, setting up a basic incident command board with initial assignments, or operating a pump panel. Then develop more intricate rehearsals of searching a bedroom or disentangling from wires. The goal is to make the rehearsal realistic, vivid, and memorable by involving more of the senses and increasing the degree of difficulty a little at a time. Only after you become competent should you add complications or troubleshooting plans.

You must have a purpose to succeed. Develop a plan with identifiable goals to be reached during the mental rehearsal. Allow your success to determine the extent of the development plan.

DEVELOPING A MENTAL REHEARSAL PLAN

Like a good preplan of a building on fire, you must initially assume the role of a story designer, which is similar to a movie director designing a sequence of events for a car chase in a movie. Develop a short storyboard of what actions will happen and in what order. Determine what is most appropriate and realistic for an effective resolution to the incident.

Moving into a mental rehearsal exercise can be easy (fig. 7–9). Without proper training, however, it can spin out of control and lack direction. To be successful building a storyboard of actions, you need focus and a dedicated time for planning. Choose a quiet area free of distractions to create your plan.

For purposes of building skills and developing a mental rehearsal plan, it is important to start with something you can easily visualize in your mind's eye. Choose a simple skill that has been frequently repeated and is second nature to think through. Start with a basic 3 to 5 point rehearsal. As your competency with visualizing the mental rehearsal grows through cycles of repetition, slowly enhance it by adding more points of detail. Eventually, this could expand to 20 points or more.

Fig. 7–9. Like an SCBA use rehearsal, a firefighter can also do a mental rehearsal of a pump panel operation at the start of a shift.
Courtesy: Dave Gillespie.

Example: SCBA donning and activating the PASS

Use a shorter time frame (one minute or less) and play out the action and preferred outcome.

Imagine yourself wearing your SCBA:

- Hear the click of the diaphragm and the movement of air into your face mask.
- Feel the rush of cool air as you inhale.
- Test the emergency button for your PASS alarm. Reach up with one hand and find the button, pressing it firmly. Notice the way the SCBA mask feels on your head, the pressure necessary to press the alarm button, the piercing sound it makes, and the smell of the rubber from the mask.

Next, graduate to a longer replay with more detail, such as up to two minutes.

Example: SCBA, entanglement, and rescue

- Imagine yourself wearing the SCBA as you exit the truck.
- Hear the click of the diaphragm and the movement of air into your face mask.
- Feel the rush of cool air as you inhale.
- See yourself in a room as your legs become entangled. Call a Mayday and attempt self-rescue. Reach for the PASS alarm and activate it (fig. 7–10).
- The rapid intervention team (RIT) finds you and untangles the wires. Everyone successfully exits together.
- Again, involve all your senses. Notice the pressure points of the SCBA mask on your head, the words calling the Mayday. Feel the necessary pressure to press the alarm button, hear the piercing sound it makes, and smell the rubber from the mask.

Building Mental Toughness

Fig. 7–10. A firefighter practices how to activate a PASS alarm. *Courtesy:* Dave Gillespie.

Example: SCBA, entanglement, and rescue in greater detail

- See yourself riding in the jump seat, tightening the straps on your SCBA. See yourself practicing tactical breathing exercises (as discussed in chapter 9).
- Imagine yourself wearing the SCBA walking from the truck. Smell a hint of smoke in the air.
- Feel the wheel on the cylinder as your hand is turning on the air cylinder.
- Arrive at the door wearing your face mask. Hook up the regulator and take an initial breath.
- See the HUD light up.
- Hear the click of the diaphragm and the sound of air moving into your face mask.

91

- Smell the fresh air and feel the rush of cool air as you inhale.
- See the dark hallway with limited visibility.
- On entry, follow the proper search pattern.
- See yourself as your legs become entangled. You attempt self-rescue unsuccessfully. You call out to your crew, but they cannot hear you.
- Call a Mayday according to your policy and procedures.
- Reach for your PASS alarm and activate it.
- Either the RIT or your crew finds you and successfully rescues you. Everyone exits together.

TIMING A MENTAL REHEARSAL PLAN

Allow the first rehearsal to flow for 10 to 30 seconds. You will find it easy to reset and follow the same pattern with minor adjustments. By tweaking short mental rehearsals, you can repeat and solidify preferred actions, locking them down. Then you can expand to a more comprehensive version in subsequent rehearsals that can grow to 60 seconds, 120 seconds, and progressively longer times as your mental skills improve.

Progressive runs will slowly increase in time, eventually moving from repeated 30-second clips to more advanced, longer clips.

As you gain confidence in the process and reset the mental rehearsal, adjust each evolution. Strive to make it more complicated by front-loading the mental rehearsal and determine the following:

1. What route you will travel from the station to the scene? Visualize well-recognized landmarks (restaurants, churches, car lots, parks, etc.).
2. How will you arrive on scene? Are you staged back and upwind from the fire, directly in front, or across the street out of the collapse zone?
3. From which side of the truck will you exit? Determine your approach to the building and put tools into action.

DYNAMIC IMAGERY

Sometimes people think that mental rehearsal is all about visualizing, or seeing what happens while they sit still with their eyes closed. One of the most effective means for training your body for clear direct action is to dynamically move your body to what your mind is rehearsing.

One of best examples of dynamic imagery is that of the US Navy Flight Demonstration Team, the Blue Angels. The Blue Angels squadron commander and pilots perform aerial maneuvers, diamond formations, barrel rolls, and loops in six McDonnell Douglas F/A-18 Hornets at speeds ranging from 120 mph up to 700 mph.[11] Their mental rehearsal and ground practice prepare them for live practice with fast rolls, slow rolls, close passes, and mirror formations, such as back to back or belly to belly. At times they may fly as close as 18 inches apart (fig. 7–11).

Fig. 7–11. The Blue Angels pilots rehearse their coordinated maneuvers over their base in Pensacola, Florida before each show while flying parallel, only 18 inches apart.
Courtesy: **US Navy.**

On the ground they develop a choreographed plan of moves, positions within formations, and a sequence of tilts, leans, and banks. Before they begin practicing new choreography in the air, the pilots do a sit-down rehearsal in chairs in their training room and conduct numerous dry runs.

The team sits in a circle, with pilot and copilot sitting next to each other, eyes closed and hands out front as if on a joystick and on the cockpit controls. As the commander gives verbal directions, they acknowledge and perform the physical move by moving their hands and joystick back, tilting their body, or leaning side to side. This form of dynamic imagery is used to train the brain and the body to perform live maneuvers in the sky exactly as rehearsed on the ground.

Similarly, there are many skills a firefighter can rehearse independently or with their crew using dynamic imagery to develop muscle memory:

- Donning the SCBA
- Forcing an inward swinging door
- Hooking up the extrication tools
- Landmarking and initiating CPR
- Donning a rescue suit and life jacket for water rescue
- Operating controls to the aerial ladder
- Grabbing an extension ladder with a partner (fig. 7–12)
- Preparing an intravenous drip
- Doing a final check of the harness equipment for rappelling off a building

Fig. 7–12. A firefighter practices dynamic rehearsal of lifting an extension ladder. *Courtesy:* Ric Jorge.

CHANGING PERSPECTIVE BASED ON ROLES

A mental rehearsal plan for an engineer might include visualizing driving in a calm, controlled manner to the scene, positioning the pumper, putting the pump into gear, and charging a 1¾-inch initial attack line. A company officer's mental rehearsal could include visualizing himself or herself in the engine front seat, pulling up in front, completing a size-up, communicating an initial report with the strategy, and advising the crew of the next actions (fig. 7–13).

Fig. 7–13. A company officer practices making an initial report while sitting in the truck in the station.
Courtesy: Dave Gillespie.

A firefighter's mental rehearsal plan may involve pulling up to a working structure fire with heavy, dark smoke, developing a momentary mental image of the possible floor plan based on the windows, doors, and preexisting knowledge, and stretching an attack line to the front door.

Moving to these more advanced dynamics in fire control or rescue scenarios can make the rehearsal last up to three minutes or longer.

REPEATED PRACTICE

Wrapping up a mental rehearsal session must finish on a high note of successful completion. It will take repeated practice for this to work effectively, returning to the same mental rehearsal as needed to see the preferred outcome. If you run out of time, it is best to fast-forward and jump to the successful conclusion. You will remember the satisfaction from a well-executed mental rehearsal, which will better prepare you for the real thing.

NOTES

1. Bob Bowman and Charles Butler, *The Golden Rules: Finding World-Class Excellence in Your Life and Work* (New York: St. Martin's Press, 2016).
2. Noa Kageyama, "Dynamic Imagery: A More Effective Way to Do Mental Rehearsal?" *Bulletproof Musician*, https://bulletproofmusician.com/why-do-we-have-to-sit-still-when-doing-mental-imagery/.
3. Alyssa Roenigk, "Lotus Pose on Two," *ESPN.com* (August 21, 2013), http://www.espn.com/nfl/story/_/id/9581925/seattle-seahawks-use-unusual-techniques-practice-espn-magazine.
4. Bob Bowman and Charles Butler, *The Golden Rules: 10 Steps to World-Class Excellence in Your Life and Work* (Little, Brown Book Group, 2016); Anand Damani, "How Michael Phelps' Coach Trained Him," *Behavioural Design* (July 10, 2013), http://www.behaviouraldesign.com/2013/10/07/how-michael-phelps-coach-trained-him/#sthash.gboFyHFZ.xr0euXFa.dpbs.
5. Judith Andersen, Marian Pitel, Ashini Weerasinghe, and Konstantinos Papazoglou, "Highly Realistic Scenario Based Training Simulates the Psychophysiology of Real World Use of Force Encounters: Implications for Improved Police Officer Performance," *Journal of Law Enforcement* 5, no. 4 (2015).
6. Ibid.
7. Donald Meichenbaum, *Cognitive-Behavior Modification: An Integrative Approach.* (Waterloo, Ontario: University of Waterloo, Springer Science Business Media, 1977).
8. Dave Grossman and Loren Christensen, *On Combat, The Psychology and Physiology of Deadly Conflict in War and in Peace*, 3rd ed. (Millstadt, IL: Warrior Science Publications, 2008).
9. Michael Asken, *Warrior Mindset: Mental Toughness Skills for a Nation's Peacekeepers* (Millstadt, IL: Warrior Science Publications, 2010).
10. Nichola Callow, Ross Roberts, Lew Hardy, Dan Jiang, and Martin Gareth Edwards, "Performance Improvements from Imagery: Evidence That Internal Visual Imagery Is Superior to External Visual Imagery for Slalom Performance," *Frontiers in Human Neuroscience* (October 21, 2013), https://www.frontiersin.org/articles/10.3389/fnhum.2013.00697/full.
11. "Visualization and Imagery—Military," APG Services, US Navy Blue Angels in Rehearsal (2012), https://www.youtube.com/watch?v=3ZNnaXj9SvY.

8

The Power of Self-Talk

SELF-TALK

From childhood onward, self-talk can either enhance or complicate our lives. We know from cognitive researchers that the rate of internal dialogue typically varies between 300 and 1,000 words per minute. When a firefighter has a less-than-ideal performance at a fire scene, or makes a mistake, it is common for him or her to be self-critical with internal talk. If this negative self-talk continues, it can become rumination. It is vital that we take control of our self-talk and ensure it is positive and constructive. Positive self-talk will allow us to become successful in high-performance actions on a consistent basis.[1]

THINKING TRAPS

Taking control of your self-talk will be the most challenging fight you will encounter. Negative thought patterns can be developed from an early age, and they can also be the result of a traumatic event. Persistent self-talk that goes unchecked can easily get pulled into a negative pattern of internal dialogue. This type of thought process is what we call *thinking traps*, or negative thoughts based in fear.

Some examples of thinking traps are as follows:

- Focusing on discomfort = physical (restrictive movement, heat, sweating)
- Focusing on things out of my control = mental (policies, studies and observation groups [SOGs], others' opinions and actions)
- Intense emotions based on repetitive thoughts or previous experiences = mental and tactical (not being properly prepared through training or otherwise)

- Distressing thoughts = mental and emotional (running out of air, getting burned, dying)
- Black-and-white thinking or extreme thinking = mindfulness (success or failure with no possibility of anything between)

To overcome thinking traps, you must first become aware of them. Coworkers might point these traps out to you voluntarily, or you could ask them for help. Humility allows for improvement, and it keeps us on an even keel. Once you identify traps, you can challenge these thought patterns or beliefs. Confusion is a typical by-product of challenging old beliefs, and you should not allow it to deter you from changing.

Familiarity breeds comfort, but do not allow familiarity to come as a result of poor training, inaccurate thought processing, or bad judgment based on false information. Whether a person works in the fire or police service, or as a paramedic or accountant, it is very normal to have periods of negative self-talk. It is common and can be part of a standard day on the job or at home.

The problem is when we allow it to perpetuate, eventually affecting other unrelated events and thoughts. This is where we must have a system to acknowledge the negative internal dialogue and stop it. This can be done by recognizing and using verbal key words or key phrases. Keeping it to three to five words will have more impact, such as "Stop the crap talk." Addressing yourself as *You* instead of *I* will be more effective because you are coaching yourself. For example, you might say to yourself, "Hey Dawson, you have got to stop the crap talk." Results have shown that people were more positive and less emotional when having a constructive self-talk internal dialogue.

Allowing negative self-talk to continue will degrade the original objective and may hinder any potential success. Overcoming this trap, however, will require a systematic effort to acknowledge the negative internal dialogue and abruptly halt it.

Some examples of self-statements used by high performers to stop negative thoughts include the following:

- "Stop the talk. You are not going to the dark side."
- "New direction. You can make this work."
- "Time to re-jig. You are in control here."
- "It'll buff out."

When a person experiences negative self-talk and it grows at an accelerating rate, it begins to have a physiological impact. Effects can include the heart beating faster, hands shaking, or perspiration. Left unchecked, it can become an anxiety attack, also known as a *panic attack*. Effectively the person has triggered a sympathetic nervous system stress response, which can impact his or her immediate situation. It may range from stuttering to doubting if they are right, not moving, choosing the wrong route down the hallway, or freezing up on the radio.

Once we know our triggers and common self-talk practices that can hurt our performance, we can label the effect they are having. This is called *effect labeling*. Once an effect is labeled, a person tends to have more control in reducing its presence, and therefore its impact. Once you recognize your internal dialogue and the triggers it has on your physiology, you can use the opportunity to change your self-talk and work to condition your thinking to more positive cues.

TACTIC #2: CONSTRUCTIVE SELF-TALK

The use of constructive self-talk is considered the most readily available performance tool to enhance actions prior to a call on the training ground and during an actual incident.

Most people seen talking to themselves are the focus of jokes, but in truth, we all do it, says Dr. Michael Asken, a psychologist for Pennsylvania State Police and Tactical Teams.[2] Most of us do it internally and are unaware of it. With proper instruction, self-talk can turn an unsure lieutenant or unconfident rookie into a powerful firefighter. Empowering yourself can lead to empowering the crew.

Constructive self-talk is used internally to aid in promoting a strong command and control disposition, good decision-making, and calm radio communications. Overall, it leads to better communications and a calmer state of mind. This can make the difference between a line-of-duty death or saving your own life and your partner's life. Most people who establish a positive mindset maintain a positive attitude. This leads us to assume their self-talk must also be positive. As Dr. Asken points out, self-talk always occurs before you act or say anything. Self-talk occurs rapidly. Cognitive psychologists estimate that an average person who conducts self-talk can have an internal dialogue with as many as 300–1,000 words per minute.

Research has determined that humans can listen and comprehend 300–400 words per minute (typical speaking is 100–150 words per minute). This means our mental processes and self-talk can exceed normal conversation and fill our minds with related and unrelated distractions.[3] Once you become aware of your negative self-talk, you can change it. With practice, the nature of self-talk can be refined to be quick, quiet, and constructive.

Self-talk is more than just saying, "I think I can...I think I can," says Lieutenant Colonel David Grossman, former US Army Ranger, psychology professor from West Point, and trainer and speaker to elite special forces and SWAT teams. Lieutenant Colonel Grossman points out our self-talk can be influenced by how we see ourselves and what we tell ourselves about us. He notes, "One Vietnam veteran, an old retired Colonel, once said this to me: 'Most of the people in our society are sheep. They are kind, gentle, productive creatures who can only hurt one another by accident.'" Like sheep, they go about their daily lives in peace for the most part. Grossman says that the colonel added, "'Then there are the sheep dogs...and I'm a sheepdog. I live to protect the flock and confront the wolf.'"[4] This type of self-talk empowers the emergency responder with mental images and motivation to act with purpose and protective instincts.

If the firefighter has an internal dialogue of 300 to 1,000 words per minute in the station house, or during emergency response, then what we tell ourselves on the job is more important than what we verbalize to others.

An old-school training session at a firehouse might have included comments such as, "Tough crap, Rookie—deal with it," "Keep your head on," or "Don't screw up!" More progressive training instructors promote the use of terminology that is constructive and direct. Constructive self-talk is choosing a line of words that motivates us to perform at our highest capacity. The relevance of cue words or performance phrases is often overlooked.

Table 8–1. Words for mental toughness in the fire service

LESS EFFECTIVE WORDS	MORE EFFECTIVE WORDS
Why am I doing this?	You got this.
	This is an opportunity to test my skills.
This is tough.	You have experienced "tough" before and have gotten through it.
	I am good at this./You are good at this.
Holy <expletive>!	All right—slow down, breathe, stay focused, and come up with your plan.
	Bring it on!
What's going to happen?	You are a professional.
	You are trained for this.

We can better prepare ourselves when focusing on constructive self-talk that supports preferred outcomes.

FIVE TYPES OF SELF-TALK IN THE FIRE SERVICE

When faced with high-risk, high-stress rescue operations, self-talk can be divided into five categories:

1. General self-talk of the operation
2. Motivational self-talk
3. Task-relevant instructional self-talk
4. Unrelated self-talk
5. Reactionary self-talk

Thoughts that are tied to activities at home, involving tasks not present, can hinder your concentration on the upcoming operation. To change from being preoccupied, it is necessary to initiate arousal control strategies.

1. General self-talk of the operation

The first type of self-talk focuses on the anticipated fire or rescue operation, including the sequence of actions at the scene. In the fire service, a crew analyzes the potential for a fire or accident and lists what units will respond, where they would stage operations, and what the priorities and initial action plan will be for water supply, initial attack, rescue profile, and addressing hazards. This general self-talk of the operation highlights the sequence of various actions that come together to make the team function.

2. Motivational self-talk

This method of constructive self-talk is to inspire yourself to maintain the effort until completing a task. Whether it is running a marathon, searching a house, or preparing to establish command at a car crash or commercial fire, motivational self-talk is effective in keeping a positive outlook on the upcoming task. In a world with constant distractions and detractors, we must be our own champions to get things done successfully.

It is best to start with a simple phrase and then develop a series of memorable phrases as you develop a positive internal dialogue. Examples of simple phrases could include the following: "This is totally in my ability," "You got this 100%," or "This is my specialty."

Fig. 8–1. A three-person fire crew utilizes self-talk before a rescue operation. *Courtesy:* Dave Gillespie.

3. Instructional self-talk

This method relies on using key descriptive words that help you complete the task at hand by instructing yourself what to do and in what order.

An example is when forcing a steel entry door. You may perform self-talk with phrases such as the following: "Shock the door, gap the door at the frame, place Halligan forks in for striking, then work the shaft to lever open the door." (See fig. 8–2.)

4. Unrelated self-talk

This is the category of useless things we think of at inappropriate times. Unrelated thoughts direct our internal voice and mental focus somewhere else, even if briefly. This contributes to short gaps in concentration and creates problems. Unrelated self-talk might occur when you are responding to a car accident while replaying in your mind the words your spouse said last night. This might also occur if you are making a mental list of the groceries needed for the weekend while attending a training session.

Fig. 8–2. A fire team prepares themselves with instructional self-talk prior to forcing impact glass during training.
Courtesy: Dave Gillespie.

An example of unrelated self-talk might include a situation in which you are en route to a call, and your mind is directed elsewhere: "How am I going to cover the excessive cost for the new addition at the house?" "Should I go for that promotion?" "I'd make a better captain than the last guy." "What do I need at the grocery store?"

5. Reactionary self-talk

Reactionary self-talk takes place after an event or action, such as swinging a golf club. If you swing accurately and hit the golf ball as intended, "Congratulations!" may be part of the internal dialogue. If you miss the ball, the conversation sounds very different. We have all been there when someone on the crew has missed a detail and others have reacted to it. Perhaps someone backed the rig into a civilian's car or handled the nozzle poorly. Self-talk after the event can set your mood for the rest of the day.

As an example, reactionary self-talk might occur if an engineer misses a street and drives by the address. The engineer might respond with reactionary internal dialogue such as, "I missed the cross street and now we are delayed by two minutes. How could I have screwed that up?"

DEVELOPING AN ANCHOR PHRASE

An *anchor phrase* is an easily recalled string of words that can be used to promote the dumping of excess thoughts and emotional baggage. It facilitates the return of focus to a preestablished mind-set, state of ease, or state of motivation. Examples of such phrases include the following:

- I will live through this.
- This is my day.
- Now I run to the problem.

One creed used by Navy SEALs to prepare for hard effort is the phrase, "The only easy day was yesterday."

Fig. 8–3. An example of positive self-talk using an anchor phrase
Photo credit: Mark VonAppen, Fully Involved.

The SEALS' anchor phrase reminds the person in dialogue that yesterday was easy, and tomorrow, we will see today as easy. The task may seem overwhelming, but it is a doable task. "I will get through this and thrive" (see fig. 8–3).

Examples of anchor phrases used by current firefighters interviewed include the following:

- Just breathe.
- I am in command of my situation.

- I embrace problems.
- Let's keep this from getting worse.
- Let's make their day better.
- Run to the fire. Whatever that "fire" is, I face it head on.

UTILIZING SELF-TALK AS A PROACTIVE TOOL

It is critical to choose the type of self-talk that happens before a situation unfolds. It must be specific and productive to help enter in the event with the preferred circumstances. It can be referred to as a mental checklist of things to do.

Example of proactive self-talk

On route to a house fire, an engineer may be mentally rehearsing upcoming actions, such as, "Go down 2nd Avenue, turn left by the big church. Continue cautiously because there will be kids in the street. Turn right onto Waterford Street and start searching for #595."

In the back seat, a firefighter may have internal dialogue such as, "If the house is on my side, grab the nozzle, then the first loop of the preconnect. Run to the front door of the house."

The company officer's internal dialogue might include thoughts such as, "It's day time, so people may be home. Watch out for the kids playing near the street. Get my three-sided view and take a tactical breath cycle before transmitting on the radio."

Developing an arousal control image or phrase

The fire service is in an excellent position to adopt images to help control our emotions and state of arousal. We have seen many examples of civilians who stress out at small incidents. When a firefighter brings a solution-making mentality to the situation, people tend to calm down and trust that it will work out. Similarly, firefighters can use phrases and images to control their own emotions.

Arousal control phrases. Examples of arousal control phrases include the following:

- Slow down. Just breathe. Take a deep breath.
- Close your eyes. Shake it out.

Arousal control images. Examples of arousal control images or sounds include the following:

- A favorite calming spot like a beach or camp

- The sound of waves, or the sound of wind through the trees
- A face of a close person or animal that is calming

These images can be uploaded as screensavers on your smartphone.

Dr. Asken states that training instructors can influence this process by replacing phrases related to what to do *if* a situation occurs with training for what to do *when* this situation occurs.[5] This changes it from being a possible situation to a probable situation. The probability may be low, but it exists. Therefore, having a contingency plan for your toolbox minimizes the surprise and raises the level of professional response.

COPING STATEMENTS

Patterns require repetition to develop, and it will require even more repetition to alter them. Neural pathways require many repetitions before an action, thought, or process becomes automatic. To date there is no hard, researchable evidence that identifies a set number of repetitions necessary in order to develop automaticity. But it is understood that the number of repetitions needed to overcome a learned action, thought, or process is greater than the number of necessary repetitions to develop automaticity of a new action, thought, or process.

Motor memory is often used interchangeably with *muscle memory*, but they are not the same. To develop the necessary motor memory to overcome or unlearn a process or series of actions that are no longer functional or accurate requires repetitive work. Contrary to popular belief that you can't teach an old dog new tricks, it *is* possible to retrain someone to new coping statements and habits. Often people quit trying to change, thinking they cannot grasp the process or that it does not work. In reality, change is possible, but people need to understand the laws of learning and how their minds work. The deeper and more ingrained a mind's neural pathway development, the more work will be required to overcome it. Often it took years to develop the fear and the pattern of response, and it will require time and diligence to overcome them. While the process is not easy, understanding how the mind works and developing coping statements can help tremendously.

Continued training in self-talk increases confidence

Coping statements are typically the antithesis of the problem. If my self-talk is negative, then I must turn it into positive self-talk. The goal is to develop weaknesses into strengths. Thinking traps are fought in the mind, so preparing statements ahead of time that are reinforced repetitively in training will work best.

An example of this process could be as follows:

Problem: "I get claustrophobic sometimes, and my mind tells me I can't breathe."

Solution: Use a gradual form of training in which the goal is to develop comfort in small incremental gains. Combined with mindfulness, the thought of not being able to breathe can be overcome.

Mindfulness will allow you to stay in the moment, and it is after the small successes that your coping statements change, such as, "I have done this before" or "I can do this, I trained for it." In this example, the coping statements that were developed are reinforced with actual training to overcome a weakness.

By continued training in constructive self-talk, and gradual increasing your use of cue words and short phrases in difficult situations, you will gain greater confidence.

TACTIC #3: GOAL SETTING AND SEGMENTATION

Firefighters have varied backgrounds and achievements as athletes, military service members, laborers with specific skills, mechanics, and technical specialists. All have proven they can accomplish goals and demonstrate commitment.

When someone is hired in emergency services, it is often because that person has demonstrated passion, technical competency, and an ability to achieve goals. However, it is rare that these people have analyzed the methodology they use to achieve success. They simply follow the goal-setting patterns set out in earlier years from school, sports, role models, or employment. With that lack of formal training, it is no surprise that some people only partially succeed in reaching more advanced or challenging goals.

In contrast, high-performance people develop formal strategies with specific tactics that work for them. High-performance firefighters apply these strategies to achieve goals related to their careers, fitness and health, financial success, life experience, or firefighting skills or advancement.

Firefighters are often viewed by the public as mentally strong men and women with strong goal-setting skills, people who complete a higher level of daily goals than the standard civilian. The only difference between those in the fire service and the average civilian is that firefighters are surrounded by goal-driven people who are preparing to save lives and make a difference.

TYPES OF GOALS

To understand how we set goals, good or bad, or how we succeed or fail, it is best that we comprehend the five types of goals:

- Subjective goals
- Objective goals
- Outcome goals
- Performance goals
- Process goals

Subjective goals

Subjective goals are related to a general task, not a specific performance level. A person who sets a subjective goal simply states that they will apply effort and try their best, without committing to specific expectations or outcome.

Examples of subjective goals. Examples of subjective goals include the following:

- General subjective goals:

    I will try to lose weight.

    My goal is to run in a road race.

- Subjective goals for a firefighter:

    My goal is to get out of the station faster.

    My goal is to don my gear faster.

Objective goals

Objective goals are specific in that they are based on a firefighter's performance on a given event or with a given task. They may be run as an individual activity or in sequence.

Examples of objective goals. Examples of objective goals include the following:

- Objective goal for an athlete: To complete a three-mile run in 25 minutes.
- Objective goal for a firefighter: To decrease turnout time by 20 seconds by next Tuesday's shift.

Outcome goals

Outcome goals are related to general outcomes to be achieved.

Examples of outcome goals. Examples of outcome goals for an incident commander, an athlete, and a firefighter are as follows:

- Outcome goal for an incident commander: To have the crew extinguish the house fire.
- Outcome goal for athlete: To place in the top five at next month's race.
- Outcome goal for firefighter: To get dressed and on the truck before anyone else for the next call.

Performance goals

Performance goals are related to multiple tasks using measurable statistics that contribute to overall improvement in job performance (see fig. 8–4).

Fig. 8–4. A crew sets a performance goal of getting dressed, on board, and seat-belted in 60 seconds.
Courtesy: David Gillespie.

Examples of performance goals. Examples of performance goals for a basketball player and for a firefighter are as follows:

- Performance goal for a basketball player: To sink 75% of free throws.

- Performance goal for a firefighter: To reduce turnout from 120 seconds to 80 seconds on 90% of our engine calls by next month (i.e., *NFPA 1710* [Career Standard] or *NFPA 1720* [Volunteer Standard]).[6]

Process goals

Process goals are similar to, and related to, performance goals. They are a series of mini-actions that, when taken in sequence, contribute to the performance goal.

Example of process goals for a hydrant catch include the following:

- Grab the hydrant line and wrench, position the hose against the hydrant.
- Drop to my knees and apply wrench to the hydrant cap.
- Match the threads and spin the coupling to tighten the coupling.
- Put the wrench on the stem and signal to the pump operator.

"Athletes often set goals that are not specific or measurable," says Dr. Terry Orlick.[7] Athletes who are naive often set goals that focus on winning, but they may have little control over whether they win. This blind targeting of goals frequently leads to failure or only partial success. Ultimately if we are to be effective, we must define a specific, performance-oriented goal, with criteria to measure its achievement.

SMART GOALS

A SMART goal is derived from a standard acronym to remember key components of effective goal setting. It reminds us that we are capable of setting ourselves up for success if we can follow a structured format and take one step at a time. The acronym SMART can be useful in goal-setting exercises:

Specific:	What is the detailed task?
Measurable:	Can we set a benchmark to determine if it is achieved?
Attainable:	Is the goal reasonable to attain?
Realistic:	Is the goal realistic given our physical attributes, location, and situation?
Time-sensitive:	What is the timeline, date, or critical point at which we will evaluate whether or not we reached our goal?

DESIGNING SMART GOALS

The use of preferred process goals and performance goals with SMART components makes it more likely that targets will be achieved. This is true whether the target is attaining functional fitness, eating healthier, stretching a hoseline, performing a rigging drill, inserting an oropharyngeal airway, or forcing a prop door. The SMART format is easily applied to emergency and nonemergency goal-setting exercises, but care must be taken not to misapply it. There are times when the mission is overwhelming, and larger goals must be reduced in scope. In such cases, it will be easier to accomplish smaller goals in order to eventually achieve larger goals. The challenge can be with mission-critical emergency operations such as a house fire, multivehicle accident, or water rescue, where response time is of the essence.

SEGMENTING A TASK WHEN IN CRISIS

The principle of segmenting a task is another concept that has transferred from athletics, special forces, and tactical police team training to firefighting.[8] *Segmenting* or *chunking* is the process of breaking down a larger goal or a series of actions into smaller chunks. Depending on the severity of the situation and the need for high performance, those physical actions may need to be grouped in a simple package of doable tasks. In the case of firefighting and rescue work, the goal setting can be applied to both administrative and operational situations.

On a large-scale operational basis, firefighters apply segmenting within an incident management system, and teams are tasked to complete certain tactics. The goal may be to provide water supply, initiate a fire attack, or vent a roof, but the means to accomplish the goal is through reliance on division of labor. By dividing a major event into sectors or divisions, functional tactics, or geographic zones, crews are assigned to focus on one component of the operation in a manner that basically segments the incident into smaller areas of response.

Likewise, on a personal level, segmenting can be an effective technique for an individual who learns to divide larger goals into more bite-sized tasks that are less overwhelming. Segmenting can be useful to change the perception that the event or obstacle is impossible by dividing it into more doable tasks. When a firefighter routinely applies this principle to overwhelming tasks, he or she develops a reputation for being a high-performance, get-it-done individual.

Example of segmenting a task on the fireground. A fireground example of this could occur in a situation in which a firefighter is entangled in wires during a response to a commercial fire. The goal may be to exit the building, but the process of disentanglement from the wires requires that the stressed-out firefighter segment the process into smaller steps as follows:

Situation: Dang, I'm stuck in these <expletive> wires!

Goal: I must get out of these wires to exit safely.

Segmenting the goal:

1. Segment 1: Control my heart rate.
   - Stop moving.
   - Breathe. Slow my heart down. Control my breathing.
2. Segment 2: Call for help.
   - Call for my team by voice or radio.
   - Initiate radio communication to IC, advising that I am tangled in wires and attempting to self-extricate.
   - I can activate Mayday, if needed. If appropriate, find my PASS and hit the button.

Fig. 8–5. A firefighter segments goals to 1) call for help, 2) reach for his tools, and 3) self-extricate from downed wires.
Courtesy: Central York Fire Services.

3. Segment 3: Disentangle.
   - Initiate simple move to disentangle.
   - Reach for cutting tool in my left pocket.
   - Grasp wires and start cutting (see fig. 8–5).

Segmenting a task in a crisis situation revolves around determining what you need to do to survive the next minute. Then you can focus on the next 30 seconds, and the next 30 seconds, and so on. It might grow to focusing on surviving until you reach the next room, then surviving to the next floor, and eventually surviving until you are out the door.

FREEZING

The action (or inaction) of freezing can be the result of not breaking tasks down sufficiently to make them manageable. Someone who is overwhelmed by the sheer volume or difficulty of the tasks necessary to perform a self-rescue may freeze or shut down and simply give up.

When people are frozen by indecision, or by analyzing a situation for too long, they often can overcome this response by asking themselves, "What is the next physical thing I have do?" Choosing the next physical task is a means of segmenting their response.

Example of freezing in wire entanglement scenario. In the example given previously, in which a firefighter is entangled in wires while responding to a commercial fire, the firefighter could use the following steps to move toward self-rescue:

1. Ask the question, "What is the next physical task?"
2. The firefighter gives self-instruction in response, such as, "Move my right hand to my pocket, and grasp my personal wire cutters."
3. The firefighter grasps the wires and begins to work toward self-rescue.

Running a marathon is never about running 26.2 miles. It is all about running to the next water station, the next mile, or even the next telephone pole. Segmenting a marathon is like segmenting a tough fire.

As crisis hits on the fire scene, such as the ceiling collapses or the conditions in the room approach a flashover point, our human stress response is activated. Even so, we can manage the chaos and survive. We can rise to the occasion by segmenting our response into smaller tasks and asking what physical action is required next. As we successfully segment and complete these microtasks, they accumulate into completing an action, which leads to completing a tactic. Eventually the entire goal is achieved.

TRAIN, PREPARE, AND THRIVE

Sometimes firefighters set a goal without any benchmark, measuring stick, or timeline. Simply saying they will accomplish a task by focusing on it misses out on the process that could lead them to success. The lack of specific planning usually sets us up for good intentions but offers little momentum to hold ourselves to task and to be accountable for the results.

While setting goals is key to effective time management and resource allocation, it is easy to get stuck in a cycle of planning and not doing. Perhaps you have tried to figure out how to complete a job performance review for a difficult employee. If you have rehearsed the situation, identified possible reactions, and prepared for that person, then take the next step and book the appointment and follow through.

If we are on the fireground, the fire does not wait. We set a goal, and the IC enables the crew to conduct the primary search. The crew begins the search, but if a firefighter becomes lost while performing that primary search, our response changes. We then transition from broad goals and broad actions into a new way of thinking. Similarly, to be effective in responding to other situations, we must reset our goals and segment the physical tasks into small, easily accomplishable chunks. You must train yourself and your team for the "what ifs" and prepare for these mental tactics in order to survive and thrive in the high-performance world of the fire service.

NOTES

1. Terry Orlick, *In Pursuit of Excellence: How to Win in Sport and Life through Mental Training*, 5th ed. (Champaign, IL: Human Kinetics Publishing, 2016).
2. Michael Asken, *Warrior Mindset. Mental Toughness Skills for a Nation's Peacekeepers.* (Millstadt, IL: Warrior Science Publications, 2010).
3. David Grossman and Loren Christensen, *On Combat, The Psychology and Physiology of Deadly Conflict in War and in Peace*, 3rd ed. (Millstadt, IL: Warrior Science Publications, 2008).
4. Ibid.
5. Asken, *Warrior Mindset*.
6. NFPA, *NFPA 1710: Standard for the Organization and Deployment of Fire Suppression Operations, Emergency Medical Operations, and Special Operations to the Public by Career Fire Departments* (Quincy, MA: National Fire Protection Association, 2016); *NFPA*, NFPA 1720: Standard for the Organization and Deployment

of Fire Suppression Operations, Emergency Medical Operations and Special Operations to the Public by Volunteer Fire Departments *(Quincy, MA: National Fire Protection Association, 2014)*.
7. Orlick, *In Pursuit of Excellence.*
8. Grossman and Christensen, *On Combat*; Ethan Kross, Emma Bruehlman-Senecal, Jiyoung Park, Aleah Burson, Adrienne Dougherty, Holly Shablack, Ryan Bremner, Jason Moser, and Ozlem Ayduk, "Self-Talk as a Regulatory Mechanism: How You Do It Matters, *Journal of Personality and Social Psychology* 106, no. 2 (2014): 304–324.

PART 4: DEVELOPING RESILIENCY THROUGH PHYSIOLOGICAL TACTICS

9

Breathing and Mindfulness

THE PURPOSE of this section is to clearly define the techniques involved in the development of resiliency and the physiological needs that must be met in order to maximize the effectiveness of resiliency. In a previous chapter, we discussed utilizing resilience training along with modern technology to help maximize our training effectiveness.

Breathing falls into two categories: the autonomic and somatic nervous systems. The autonomic nervous system allows you to breathe while you sleep. The somatic nervous system allows you to consciously control your breathing.

The use of breathing techniques during periods of stress can help instill calm, as well as lower heart rate and blood pressure. Practitioners of yoga have long implemented breathing techniques to maintain control over psychological and physical arousal.[1] Research offers empirical data demonstrating that breathing has significant impact on human reactions.[2] The breathing techniques we will discuss in this chapter are 1) belly breathing, 2) the hum technique, 3) box breathing, and 4) breath and delivering air for safety and survival.

BELLY BREATHING (DIAPHRAGMATIC BREATHING)

Belly breathing, also called *diaphragmatic breathing*, is a more efficient way to breathe than chest breathing. Belly breathing allows for a complete expansion of the lungs and is proven to have homeostatic effects, which support an individual's cognitive ability during periods of stress. Chest breathing is one of the characteristics of hyperarousal of the sympathetic nervous system (SNS). During periods of high stress or anxiety, it becomes part of the physiological anxiety response.[3] Chest breathing is identified when you expand the chest wall on inspiration instead of drawing the diaphragm down into the abdomen to expand the lungs. A noticeable rise and fall of the chest, along with an absence of movement in the abdomen, indicates when a person is chest breathing. Symptoms of chest breathing include muscle fatigue (intercostal muscles), overall body tenseness, shortness of breath, and increased feelings of anxiety. Conversely, people who are belly breathing expand their lungs more fully due to the involvement of the diaphragm, thus absorbing greater amounts of oxygen due to fuller expansion of the lungs, increasing parasympathetic tone, and reducing anxiety.[4]

Belly breathing is easily learned and can be self-taught. Benefits of belly breathing include decreasing the heart rate, lowering blood pressure, and restoring cognitive function (see fig. 9–1). There are a few points to remember when learning to belly breathe:

1. Inhalation is done through the belly, which extends on inspiration.
2. On exhalation, the belly collapses.
3. The chest does not rise or fall when belly breathing.

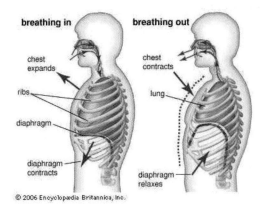

Fig. 9–1. Belly breathing is easily learned and can be self-taught.

THE HUM TECHNIQUE

The hum technique, also called the *Riley Emergency Breathing Technique*, is an effective way to extend your air. By humming on exhale, you can easily double the length of your respiratory cycle. This technique has the benefits of lowering blood pressure and heart rate, as well as extending air consumption times for SCBA use. There are a couple of points to remember when practicing the hum technique:

1. Inhale as you normally would.
2. Create a low hum, while exhaling in a slow and consistent manner.

The technique revolves around learning how to exhale while producing an audible hum. It is important not to overextend the cycle, which could cause greater fatigue. The release of air using the hum takes practice and must be mastered to develop efficiency. The exhalation must be done in such a way that it feels natural and not forced. Successful use of this technique can be greatly affected by the degree of fatigue and stress you are experiencing.

BOX BREATHING

Box breathing is also known as *square breathing* and *combat breathing*. Box breathing is a simple technique used for stress relief and coping with anxiety (see fig. 9–2). The technique follows a predetermined cadence (or count) that often starts with four seconds but can be any time cycle that is comfortable. Steps for box breathing are as follows:

1. Inhale for four seconds.
2. Hold for four seconds.
3. Exhale for six seconds.
4. Hold for two seconds.
5. Repeat the cycle.

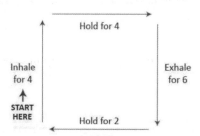

Fig. 9–2. Box breathing is also known as *combat breathing*.

Box breathing can be combined with belly breathing, which can be used for respiration efficiency. The cadence of box breathing determines the length of the cycles. When combined, these two techniques are very powerful in slowing respirations and regaining cognitive control. Box breathing has been adopted by the US Navy Development Group (DEVGRU) for their elite operators.

BREATH AND DELIVERING AIR FOR SAFETY AND SURVIVAL

This technique was developed by the authors and can extend time of air consumption significantly. The technique is predicated on several hundred trials done in training, and it has been found to work exceedingly well when done correctly. If you are lost, disoriented, low on air, trapped, or unable to exit, and you have an area of relative safety, such as a room with the door closed, this technique may increase your profile for survivability considerably.

The first step in the technique is to begin with deep belly breathing and continue until you regain composure. Next, you belly breathe deeply and hold your inhaled breath for up to 10 seconds (or whatever amount of time is comfortable for you). Following the hold, begin to exhale, producing an audible hum or hiss until your breath is completely expelled. Repeat.

The length of time you hold your breath will be different for everyone since lung capacities vary, but the idea is to begin exhaling before it becomes uncomfortable to hold your breath any longer. Exhalation is extended (just as in the hum technique). The breath should be completely exhaled without prolonging the exhale cycle unnecessarily during the hum period. It is worthy to note that in breath work, too much exertion or effort creates stress, and stress increases air consumption. The goal is a natural and relaxed technique, not a forced one.

Using this technique has allowed students to slow respirations to as low as two breaths per minute. With certain air packs using integrated pneumatic-driven safety features, you can extend "low air" significantly by combining this technique with the wheel technique. In training, the authors have witnessed low air times of 20 minutes and greater as the norm. Times ranging from 40 to 55 minutes are common. Low air times greater than these can be accomplished but are not as common.

THE WHEEL TECHNIQUE

By manipulating the cylinder valve handwheel on the SCBA to the On and Off positions, you can limit, or eliminate, the use of air for the integrated audible low-air warning safety feature. The goal is to avoid activating the low-air warning (vibration alert, bell, or whistle) and thus leave more air for breathing consumption. There is a "sweet spot" on every SCBA that allows this to occur. Finding that spot with proficiency, however, requires repetitive training under low-stress conditions (see fig. 9–3).

Breathing and Mindfulness

Fig. 9–3. Firefighter practicing breathing technique and wheel technique under low stress to improve competency.
Courtesy: Ric Jorge.

Once competence with this technique is accomplished, you can begin training under realistic conditions to gain mastery, as discussed previously in the text. The significance of using this technique is an improved survivability profile through extending your time and air supply. This increased time improves the chances of RIT reaching you before you run out of air.

Testing SCBAs by timing them from activation of the low air warning until the feature ceases to work will allow you to establish a baseline time value, and an air consumption value of each SCBA tested (see fig. 9–4).

Preliminary numbers documented by the authors have revealed a range of times from 80 to 110 minutes at the low air capacity warning. The wide range of numbers can be attributed to our industry standard for activation requirements. These are described in *NFPA 1981*, section 6.2.3, as specified in 42 CFR 84 for 2007-compliant SCBAs, which created a 20%–25% activation range. However, for 2013-compliant SCBAs, NFPA changed the design requirement in *NFPA 1981*, section 6.2.6, by requiring the alarm range for low air to be a 33% minimum.

Thus, the pressure range can vary between 1,035 and 1,215 psi (20% to 25%) just before the low air safety feature activates on the 2007-regulated model SCBAs tested. This test allows for a perspective to see how much air is being robbed from the user to activate and run the low air warning until it ceases. It stands to reason that any pneumatic safety feature would consume air that could be used by the person wearing the pack.

MINDFULNESS

Firefighting is a stressful, highly technical, and often chaotic career. Mindfulness is an excellent tool to develop a firefighter's mental preparation on and off the job. Mindfulness is a state of active and open attention to the present, a deliberate choice to live in the moment. Mindfulness is a "mental mode characterized by attention to present moment experience without conceptual elaboration or emotional reactivity. It includes the ability to pay attention to, describe and act with full awareness and sensations, perceptions, thoughts, and emotions."[5]

Mindfulness is best practiced in quiet settings, and it is best developed with frequent repetition. The following techniques to stay in the moment are simple, positive statements that will influence your successful use of mindfulness.

- The only way to improve performance is to train.
- Focus on what you can control—the present. This eliminates worrying about the future.
- Get into the "flow" and immerse yourself in what you are doing.
- Lean into fear; avoiding it feeds it.
- The humility of acknowledging that you do not know it all alleviates stress.
- Mindfulness improves your awareness.

Fig. 9–4. Practicing mindfulness helps focus attention on a critical task that may be crucial to survival such as extending available air in an SCBA.
Courtesy: Ric Jorge.

Mindfulness is the technique of thought focusing and purposefully "catching" your mind to prevent it drifting to other thoughts. Mindfulness translates into improving your focus of attention. The practice of purposefully concentrating on something, and bringing your attention back to the original thought when your mind starts to drift, improves your ability to focus. Thought focusing can be used in many different ways. For example, on your drive to work, make yourself aware of weather conditions that may have a direct impact on how you approach the job. Focus on different conditions that may influence your actions during your tour, such as the following:

- Rain (snow, sleet, flooding). You can expect slippery conditions on road ways when it rains (car accidents), as well as the possibility of flooding, possible roof collapses due to pooling water (hurricanes, strong cold fronts), snow drifts, the loss of road markers that would define a roadway, black ice.
- Wind (direction, strength). Winds associated with storms/hurricanes will determine ventilation, forcible entry, and strategy for fighting fire.
- Day of the week. Weekend, weekday.
- Holidays. For example, suicide seems to spike during the Christmas holidays.
- Special events. For example, a local sporting team wins the playoffs, which may lead to overzealous fans rioting.
- Physical health. Are you in top physical condition for the shift, or are you "dinged up"?
- Ability to focus. Are you or another crewmember experiencing physical or emotional personal issues?

Mindfulness is a form of preparing the mind for where the body may have to go. Becoming aware of the different influences, man-made or nature driven, will benefit us in mental and tactical preparation. A by-product of this technique is an increased ability to focus, calmness, and increased neuroplasticity. *Neuroplasticity* is "the brain's ability to reorganize itself by forming new neural connections throughout life."[6] Using mindfulness is a purposeful way to encourage neuroplasticity. The ability to focus and the resulting calmness developed through mindfulness training will improve our ability to perform.

OVERCOMING NEGATIVE SELF-TALK

Sometimes we lie in bed attempting to sleep, while continuously thinking about problems or upsetting or stressful situations. At other times, our minds will "run" the minute we awaken from sleep, focusing on distressing thoughts or situations. It is not uncommon for these thoughts to recur throughout the day. These are all examples of rumination.

Rumination is considered to be negative self-talk and can determine many of our outcomes before we even get started. Negative self-talk will be the most challenging fight you will encounter because it can develop from an early age or as a result of a traumatic event. This type of thought process, negative self-talk, develops thinking traps, as discussed previously. Thinking traps are negative thoughts based in fear rather than solutions.

Some examples of thinking traps:

- Focusing on physical discomfort, such as restrictive movement, heat, or sweating
- Focusing on things out of one's control, such as policies, SOGs, or the actions or opinions of others
- Intense emotions based on previous experiences, such as not being properly prepared through training
- Distressing mental and emotional thoughts, such as thinking about running out of air, getting burned, or dying
- Black-and-white thinking (or extreme thinking), such as seeing either success or failure, with no possibility of an intermediate result

To overcome thinking traps, you must first become aware of them. Others may point out thinking traps to you, or you may become aware of them by asking others for help. Humility allows for improvement, and it keeps us teachable. Once you have identified your thinking traps, you can begin to replace them with more effective thought processes. Do not be concerned if you experience some confusion initially. This is a typical by-product of challenging old thinking patterns, but don't let it deter you from changing. It is often said that familiarity breeds comfort. If that familiarity is based on poor training, inaccurate thought processing, or bad judgment based on false information, however, it becomes imperative to change accordingly.

COPING TECHNIQUES

As mentioned previously in the text, patterns, whether positive or negative, require repetition to develop, and even more repetition is required to alter them. Neural pathways require many repetitions before an action, thought, or process becomes automatic. To date there is no hard, researchable evidence that identifies a set number of repetitions before developing automaticity. It is understood that the number of repetitions necessary to overcome a learned action, thought, or process is greater than the number of repetitions necessary to develop automaticity of a new action, thought, or process.[7]

Motor memory is a term often used synonymously with muscle memory, but they are not the same. Motor memory is the totality of the development of neural pathways. To develop the necessary motor memory to overcome or unlearn a "process" that is no longer functional, or accurate, requires repetitive work. A simple example of repetitive pattern training is switching from an air pack with a regulator that comes over the right shoulder to an air pack with a regulator that comes over the left shoulder. This is significant for donning, doffing, and low-profile techniques. The length of time that this function has been repeated will determine how well developed the neural pathway is. Often people will quit trying to undertake a change in technique, thinking they cannot achieve the change or that the technique does not work. In reality, the new technique will work, but people need to understand the functioning of the mind and the laws of learning to develop new neural pathways.

The deeper and more ingrained a neural pathway development is, the more work will be required to overcome it. If it took years to develop a fear, time and diligence will be required to overcome it. To overcome a deficit like this is not easy, but understanding how the mind works and developing coping statements can help tremendously.

Coping statements are typically the antithesis of rumination. A person's negative self-talk must be developed into positive self-talk, with the overarching goal of developing weaknesses into strengths. Thinking traps are fought in the mind, so preparing statements ahead of time that are reinforced repetitively in training will work best.

Example use of coping statements

Problem: "I get claustrophobic sometimes, and my mind tells me I can't breathe."

Solution: Use a gradual form of training in which the goal is to develop comfort in small gains. Combined with mindfulness, this will allow the thought of not being able to breathe to be overcome. Mindfulness will allow you to stay in the moment, focusing on small goals designed to achieve success. It is after the small successes that you develop coping statements such as, "I have done this before" or "I know I can do this, I have trained for this successfully in the past."

In this example, the coping statements that were developed are reinforced with actual training to overcome a weakness. You establish greater confidence by continued training, with gradual increases in degree of difficulty. This strategy can be used to develop weaknesses into strengths.

NOTES

1. Waylon Lewis, "How Old Is Yoga?" *Elephant Journal* (November 18, 2009), www.elephantjournal.com/2009/11/how-old-is-yoga/.
2. "Relaxation Techniques: Breath Control Helps Quell Errant Stress Response," Harvard Health Publishing, Harvard Medical School (January 26, 2015), http://www.health.harvard.edu/mind-and-mood/relaxation-techniques-breath-control-helps-quell-errant-stress-response; Defense Centers of Excellence for Psychological Health and Traumatic Brain Injury, *Mind Body Skills for Regulating the Autonomic Nervous System*, Version 2 (June 2011), https://www.irest.us/sites/default/files/DCOE_Mind-Body%20Skills%20for%20Regulating%20the%20Autonomic%20Nervous%20System.pdf
3. Defense Centers of Excellence for Psychological Health and Traumatic Brain Injury, *Mind Body Skills*.
4. Deanne Repich, "Overcoming Concerns about Breathing," *Anxiety Relief Solutions*. National Institute of Anxiety and Stress, Inc., http://anxietyreliefsolutions.com/overcoming-concerns-about-breathing/; Defense Centers of Excellence, *Mind Body Skills*; Harvard Health Publishing, "Relaxation Techniques."
5. Alexandra B. Morrison, Merissa Goolsarran, Scott L. Rogers, and Amishi P. Jha, "Taming a Wandering Attention: Short-Form Mindfulness Training in Student Cohorts," *Frontiers of Human Neuroscience* 7 (January 6, 2014), doi: 10.3389/fnhum.2013.00897.
6. "Medical Definition of Neuroplasticity," *MedicineNet.com*, https://www.medicinenet.com/script/main/art.asp?articlekey=40362.
7. Richard A. Schmidt and Timothy D. Lee, "Principles of Practice for Learning Motor Skills: Some Implications for Practice and Instruction in Music," in *Art in Motion II: Motor Skills, Motivation, and Musical Practice*, ed. Adina Mornell (Frankfurt: Peter Lang, 2012), http://hp-research.com/publication-list, http://www.hp-research.com/sites/default/files/publications/Schmidt%20&%20Lee%20chapter%20(Mornell%20book)small.pdf.

10

The Body and Physiological Needs

THERE ARE thousands of books available on the subjects of nutrition, physical exercise, and sleep. In this chapter we want to reinforce the importance of these areas as they relate to resiliency. If you wish to research these areas further, we suggest you read several books in your specific area of interest to gather additional information.

NUTRITION

Looking at nutrition from the perspective of eating properly requires an understanding of what foods and food supplements benefit us most. Food research is the Pandora's box of the health industry.

This chapter is not a nutritional guide, but we will offer some general nutritional tips. Most importantly, you should be aware that poor nutrition can lead to fragmented memories of dreams, depression, and anxiety. It also can mimic signs of OCD and ADHD, or exacerbate existing symptoms when combined with lack of sleep.[1]

The following are general ideas. If you wish to know more, we recommend seeking out the advice of a professional nutritionist.

Simple nutrition tips

There are several tips related to nutrition for firefighters that should be kept in mind:

1. Do not be afraid of carbohydrates. Look at them as fuel. Firefighting has a high caloric burn rate.

2. Protein is important. The average person needs approximately 1.0–1.5 grams of protein per kilogram of body weight per meal.[2]
3. Some fat is beneficial. Beneficial fats can come from nuts, avocados, olives, vegetable oils, and fish (salmon, tuna) oils.
4. Stay hydrated. Drink lots of water, and consider prehydrating. Firefighting is hard, dirty work, and sweat is a by-product of this work. This is further exacerbated by being insulated in turnout gear. These conditions work against you unless you stay properly hydrated and monitor your fluid intake.
5. Replace electrolytes. Dilute sports drinks 50/50 for best balance.

Nutrient uptake will enhance recovery and performance. Avoid overeating, stay hydrated by drinking plenty of water, and make sure your body is sufficiently recovered from the previous day's activities to handle a grueling shift should the need arise.

The use of energy drinks, coffee, and alcohol can lead to an interruption in your deep sleep stage and may eliminate REM sleep patterns. These drinks can have a synergistic effect on OCD, ADHD, depression, and anxiety through continued sleep interruption.

PHYSICAL EXERCISE

The amount of energy expended when engaged in task assignments is similar between firefighters and high-level military operatives. This energy is measured in metabolic equivalents or METs. One *MET* is defined as the energy required for a healthy adult to sit quietly. Activities are then rated by multiples of that base number. Firefighting is typically considered to be in the 10 to 12 MET range.

VO_2 *max* (where V= volume, O_2 = oxygen, and max = maximum) is the maximal oxygen consumption of an individual, and it reflects his or her aerobic physical fitness. VO_2 max calculators can be found online, and establishing this level will help determine which approach to physical exercise you could take. For example, if you are strong but do not have stamina, then concentrating on your cardio may benefit you. If you have great cardio but are not very strong, then developing a strength routine would help give you the fitness balance you seek.

According to some professional evaluations, there are different areas of training that are valuable.[3] The approach to developing a balance of good health through fitness training focusing on five areas is supported by many fitness training authorities:[4]

1. Aerobic fitness
2. Strength training
3. Core exercise
4. Balance training
5. Flexibility and stretching

The approach described may be better explained in depth by contacting a local professional in your area. While the names of their five points may differ, the objective is the same: achieving a balance of training in different areas. Using dynamic, static, aerobic, and circuit training combined with flexibility exercise may create a more physically resilient firefighter.

SLEEP

Probably the most misunderstood, and most underrated, subject in the fire service could be the health benefit of sleep. Sleep is essential for a sound mind and body. Your body keeps track of the amount of sleep you get, and this concept is referred to as your *sleep bank*. A lack of sleep (on average less than 8 hours) is kept track of by your body, essentially "debiting" from your sleep bank. The research and debates about sleep reveal some common foundational knowledge about sleep. It is well known that lack of sleep influences "attention, stress, daytime sleepiness, and low-grade inflammation."[5] The latter of these symptoms, low-grade inflammation, "has been increasingly linked to a range of unhealthiness, with heart disease high on the list."[6] I will leave the prevalence of firefighter heart disease and the correlation of our sleep patterns to future research, but it certainly is worth noting.

The simplest way to explain sleep is by defining its five stages (see fig. 10–1). Stages 1 and 2 comprise the start of light sleep, transitioning into stages 3 and 4, deep sleep. In stage 1 the eyes move very slowly; muscle contraction often occurs, giving the person a sense of falling. Stage 1 is typically associated with alpha waves, as shown in figure 10–1. In stage 2 the brain and eye movement begin to slow down, and this stage is identifiable by the beginning of delta waves. The elongation of delta waves is the sign that deep sleep has been achieved and is identified as stages 3 and 4. During these stages, the body works and heals itself at the cellular level. Tissue growth and repair occur, energy is restored, and growth hormone is released during deep sleep.

Stage 5 is the REM (rapid eye movement) stage, and it is stage during which the brain heals itself. The eyes will move rapidly, breathing becomes shallow, rapid, and irregular, and heart rate and blood pressure will increase. The body becomes immobile and relaxed, as muscles seem to be "turned off." While energy is provided to both the brain and the body during REM, the cortisol levels dip at bedtime, allowing for the tired feeling. The cortisol levels increase overnight to promote alertness in the morning.

As previously mentioned in the nutrition section of this chapter, continued disruption of deep sleep cycles or deprivation of sleep can be detrimental to a person's mental and physical health.[7]

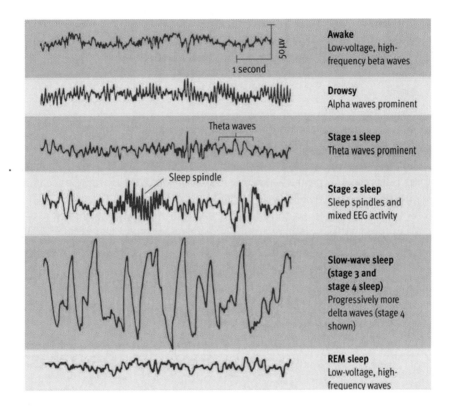

Fig. 10–1. Stages of sleep from awake to REM sleep

To catch up on lost sleep, one possibility that sleep researchers offer is the practice of sleep banking (also called *sleep binging*).[8] Sleep banking is a way to add sleep in anticipation of the upcoming loss of it. Routine naps, however, are thought to be healthier than catching up on or banking sleep. Experts refer to the effects of changing sleep habits in order to catch up on sleep as "social jet lag" because the body has become confused, as if it has traveled over time zones, interrupting the normal circadian rhythms. The recommendation of the research is that short naps (25 minutes is ideal) offer the healthiest way to catch up on sleep.[9]

Firefighters have higher-than-average rates of sleep disorders after retiring from long careers with interrupted sleep. Research in one study has shown that from a "sample of about 7,000 firefighters taken from 66 fire departments.... about 37 percent...suffered from some type of sleep disorder such as obstructive sleep apnea, shift work disorder, insomnia and restless leg syndrome."[10] The study further revealed that of those firefighters who tested positive for sleep disorders, about "80 percent . . . were untreated or undiagnosed."[11] Sleep disorders are clearly an important factor in the health of firefighters since healthy amounts of sleep can improve resiliency and response.

NOTES

1. Douglas Hunt, *What Your Doctor May Not Tell You about Anxiety, Phobias, and Panic Attacks: The All-Natural Program That Can Help You Conquer Your Fears* (New York: Warner Books, 2005).
2. Tom Venuto, *Burn the Fat, Feed the Muscle: Transform Your Body Forever Using the Secrets of the Leanest People in the World* (New York: Harmony Books, 2013).
3. "Fitness Training: Elements of a Well-Rounded Routine," Mayo Clinic, mayoclinic.org/healthy-lifestyle/fitness/in-depth/fitness-training/art-20044792?pg=1.
4. "Five Types of Fitness Training," *Livestrong.com* (September 20, 2013), livestrong.com/article/534321-five-types-of-fitness-training/.
5. David Di Salvo, "The Good and Bad News about Your Sleep Debt," *Forbes* (February 23, 2014), forbes.com/sites/daviddisalvo/2014/02/23/the-good-and-bad-news-about-your-sleep-debt/.
6. Ibid.
7. Douglas Hunt, *What Your Doctor May Not Tell You.*
8. Heidi Mitchell, "Can You Catch Up on Lost Sleep?" *Wall Street Journal* (May 20, 2013), http://www.wsj.com/articles/SB10001424127887324102604578494872502357516.
9. Ibid.
10. Sumit Passary, "Firefighters Battle Undiagnosed Sleep Disorders, Says New Study," *Tech Times* (November 14, 2014), http://www.techtimes.com/articles/20218/20141114/firefighters-battle-undiagnosed-sleep-disorders-says-new-study.htm.
11. Ibid.

11

Training Smarter

THE LEVEL of psychological and physical arousal stimulation considered appropriate for a training exercise is a subjective choice. The instructor's knowledge of these arousal levels, and the effects of adding stimulation to these levels, will determine student success or failure.

Many exercises utilized in the fire service were developed, at least in part, because of a loss of life. We approach these drills from several perspectives:

1. Are they realistic?
2. Are they functional?
3. What senses do they focus on?

Blackout exercises can be beneficial because they hinder our vision, our most dominant sense. Learning to work confidently without this sense is imperative due to conditions encountered in actual fires. There are two ways to accomplish hindering the sense of vision:

1. The student's vision can be completely obscured.
2. All light can be eliminated inside the structure being used.

Each method has advantages and disadvantages. The instructor must ensure clear goals based on the predetermined objectives. For mask confidence, the objective may simply be for the student to learn to keep calm and practice the coping techniques developed in chapter 3. If the objective is clearly stated, invaluable gains can be achieved and lessons can be learned. These lessons can progress with increasing degrees of difficulty. Using search and rescue (SAR) as an example, we can increase the degree of difficulty while monitoring the progress of individuals as they push their levels of comfort to develop greater resiliency.[1] After the students gain confidence, exercises can be strung together to create an evolution. At this point the degree of difficulty can be raised incrementally as long as the outcomes are successful. If failures become common, the exercise or evolution must be scaled back to the last point of success to reestablish confidence. The path to mastery is slow, methodical, and deliberate, but it works well.

Developing confidence by achieving the previously unachievable will increase resiliency. Resiliency training stimulates the release of dopamine, oxytocin, serotonin, and endorphins—the so-called "good" hormones. This is the very essence of success-based training discussed in this book.

Dopamine is known as the "feel-good" hormone and is associated with improving attention and cognition.[2] Oxytocin is known as the "bonding" hormone, as it enhances cooperation and trust building.[3] Serotonin is a hormone that makes you feel important by influencing "brain cells related to mood, sexual desire, appetite, sleep, memory and learning, temperature regulation, and some social behavior."[4] Endorphins reduce our perception of pain by acting as an analgesic.

BIOMETRIC TRACKING

The use of a monitoring system is necessary to fully gauge the effectiveness of training techniques. A system capable of monitoring and recording the five functions listed below is vital for success in controlling arousal levels:

- Heart rate
- Maximum heart rate capability
- Respiratory rate
- Temperature
- Heart rate variability

This system utilizes a tight-fitting shirt that has a receiving port under the left armpit (see fig. 11–1). The system used allows for real-time tracking, as well as recorded tracking of specific vitals. The information is gathered from a transmitter worn by the participant and is transmitted to a computer that is actively monitored during the exercise. The data are stored for retrieval and analysis at a later time.

Accompanied by a coaching instructor, the student progresses through a series of exercises that resemble fireground activities. The coaching instructor works directly with an individual student in this case. If the student's vitals approach levels that concern the instructor remotely monitoring the system, he or she can contact the coaching instructor via radio with this information. For example, perhaps a student's maximum heart rate capacity is

Fig. 11–1. A firefighter wears the Zephyr (hockey puck), which can record and transmit data to the command vehicle or training officer.
Courtesy: Medtronic.

exceeded for an extended period of time, and the student is not working to self-regulate. In this case, the monitoring instructor will contact the coaching instructor, who will lead the student to focus on resiliency reset techniques (such as breathing and mental rehearsal techniques). This could occur repeatedly throughout the entire process. This training process consists of a gauntlet of exercises that will tax the students, who could experience loss of small motor function and decreased cognitive ability and orientation. The coaching instructor is responsible for identifying decreased pace (student slows down, a sign of fatigue or anxiety), lack of focus or disorientation (student exhibits loss of awareness of surroundings and/or location in the building), tunnel vision, or other signs consistent with the SNS system activation.

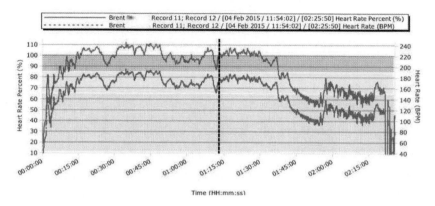

Fig. 11–2. A chart shows two tracings. The top tracing shows maximum heart rate percentage. The bottom tracing is actual heart rate.

Addressing the "extended period of time" is different for each student. Naturally a student in excellent shape versus a student who is out of shape can sustain longer periods of time exceeding his or her maximum heart rate (see fig. 11–2). The same could be said for a younger student versus an older student because stamina could become an issue. The point of using technology is to be able to give the student feedback containing very objective criteria concerning the body's response during task performance. Technology is used to map the results of the resiliency techniques during each exercise throughout the entire evolution.

Exceeding the maximum heart rate capacity quickly leads to fatigue. Left unchecked, it can render firefighters useless on the fireground, turning them into liabilities to themselves and others (see fig. 11–3). Therefore, to meet the objective, the student is only allowed to exceed capacity for short periods of time to develop physical and mental referencing for future trainings. Once students become experientially aware of these parameters, they can learn to judge for themselves how hard to push. The idea of self-awareness is not new to the fire service. Using science and technology to further individual ability creates more efficient, more productive, and more resilient firefighters.

Utilizing these techniques, the authors have conducted very productive sessions ending with debriefings in a classroom setting. Computer-generated graphs based on information gathered from the student during the evolution are analyzed and discussed. The individual students will review their progress through the exercises by reviewing their personal performance graphs that show exertion rate and physiological response to physical and mental stressors. Each exercise spike is associated with all of the five functions for monitoring.

Biometric training example

Assume the first exercise is an entanglement, and the student struggles and is coached in response to elevated heart rate or respiratory rate. The student is asked to remember several things while undergoing the exercise, including the following:

- How they felt
- How they coped or were coached
- If they could localize any physical discomfort
- If they experienced any negative self-talk

We explain to the students the senses that were the focus in each one of the exercises, and we ask how they felt before and after the resiliency reset. This allows for students to identify their pattern (their pace) and to recognize the internal warning signs they have when approaching their max output. This information will help them on the next training exercise or emergency response (see fig. 11–3).

Success can be measured by the responses as students undergo the exercises again. Sometimes providing familiarity is confused in training as a form of "cheating" or as being unrealistic. Instructors must understand the laws of adult learning, however, and that repetition is one of them. During the phase of competence (familiarity) comes confidence, and only when this is exhibited should you begin to increase the degree of difficulty in a task. With the new challenge for the student, the instructor can monitor the reaction to see if the student manages his *resiliency markers* (heat rate, respiratory rate, thought process, internal dialogue) appropriately, and determine if his or her performance reflects this. A student who is successful may be more methodical, but more importantly, the clarity and sense of purpose will be noticeable in their actions.

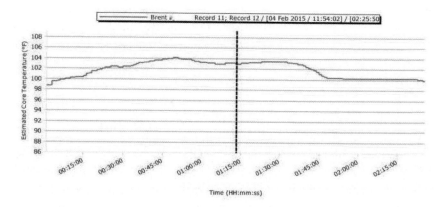

Fig. 11–3. Body core temperature shown in degrees Fahrenheit over time

THE ROLE OF BRAIN WAVES ON ACTION AND REST

Electroencephalograms (EEGs) are used to determine electrical activity in the brain. This activity is categorized in waves, which have different electrical currents that are measured in hertz. Figure 11–4 shows examples of these classifications.

These waves can tell a lot about the thinking process, which in turn has the potential to predict reactions under certain circumstances.[5] Delta waves are the slowest brain waves and can represent deep sleep. We covered some of the benefits of sleep in the sleep chapter. Theta waves are often created during meditation or insight. Alpha waves typically represent the brain resting state. They are present when you are relaxed, calm, and lucid but not really thinking (the recliner position for firefighters). Beta waves represent learning and concentration at the low end of the scale, and fear and stress at the high end of the scale. Gamma waves are fast-moving and occur when different parts of the brain are combining disparate thoughts into a single idea.

Brainwaves, Frequencies and Functions

Unconscious		Conscious		
Delta	**Theta**	**Alpha**	**Beta**	**Gamma**
0.5 – 4 Hz	4 – 8 Hz	8 – 13 Hz	13 – 30 Hz	30-42 Hz
Instinct	**Emotion**	**Consciousness**	**Thought**	**Will**
Survival Deep sleep Coma	Drives Feelings Trance Dreams	Awareness of the body Integration of feelings	Perception Concentration Mental activity	Extreme focus Energy Ecstasy

Fig. 11–4. Brain waves can inform us about relationships to our heart rate.

Understanding brain waves allows for a greater understanding of the heart rate associations with the sympathetic nervous system. EEGs can detect networks that allow us to see which parts of the brain are in communication at any given time.[6] So, when decisions are being made, EEGs can determine the structure of the network and the process used.

Two distinct systems for processing information are explicit and implicit. The *explicit system* is rule-based. It can be expressed verbally and is tied to conscious awareness. The explicit system uses the prefrontal cortex to access higher cognitive function. Using a firefighting scenario to better explain this, let's suppose you are a newly promoted officer in your fire department. You are dispatched to a structure fire. While en route, you are going through the process of remembering and gathering the information that is necessary to do the job that is required of you. The dispatcher alerts you that they are getting multiple calls on this fire and the residents are unaccounted for. The more inexperienced you are, the higher your stress. The higher the stress, the greater the likelihood you will forget something or commit errors. If you were connected to an EEG, it would clearly show a beta-wave pattern on the higher end of the beta spectrum that correlates and identifies stress or fear. Physiologically, this will show up as rapid heart rate, shallow and rapid respirations, and many other symptoms of SNS activation. This will be true of almost every learning process you go through. Of course, the stress will vary according to the nature and severity of the circumstances, but the levels are measurable nonetheless.

The *implicit system* can best be described as a gut feeling or a sense, such as a skill or experience. It is not consciously accessible and cannot be described verbally. It is like trying to explain a hunch.[7] Let's use the same scenario we used for the explicit system, but fast forward five years in time. You are dispatched to a similar call. If you have grown and improved, the thought processing is different, the preparation is different, and so is the stress level. You will be more in tune with your body's response, you will have gained clarity concerning your job responsibilities, and through repetition, you will have honed your skills. The information being forwarded by the dispatcher may not make you as uncomfortable. In fact, quite the opposite effect should occur. You will also have a better mental image of what you will be facing.

The process of creating a mental game plan began earlier in the day when you reviewed relevant factors. This could include factors such as weather conditions (seasonal challenges specific to your region), time of the year (holiday, spring break, summer time, etc.), time of the week (payday, weekend vs. weekday), and time of day (traffic density due to rush hour or holiday traffic in tourism-dominated areas). Stressors are then shifted toward the familiarity of the crew you are working with and their capabilities as individuals and a unit. What was once a very stressful event has somehow become less stressful, but it has not lost its urgency. The EEG may reveal brain waves such as low alpha/high theta waves, which is common when experience and skill meld into a career for firefighters. You will have been training diligently, and after five years in the service, you will have developed fireground skills pertinent to your position. You have honed these skills by repeated use of them under different circumstance. What was once competence became confidence and is now developing into mastery.

MASTERY

To clarify, *mastery* in the context of this book refers to training in high-stress situations. High-stress situations demand greater attention because of the risk. Risk drives focus, this increased attention improves a person's ability to more quickly assimilate information. The process of learning (competence/confidence/mastery) improves a person's ability to develop rapid pattern recognition (information chunking). This, in turn, radically increases situational awareness, as discussed previously concerning the laws of adult learning and the value of repetition.

Once achieved, mastery is not a matter of prefrontal cortex vs. amygdala, as it is during the developmental phases of mastery (competence, confidence). Instead, it is a matter of gamma waves vs. beta waves. The implicit system can include the following processes:

- Priming: Influence or previous experience
- Procedural: Repetition-induced motor memory
- Conditioning: Putting together all stimuli and responses so that they feel familiar

In the implicit system response, one of the features for firefighters is the ability to make rapid pattern recognition (wind direction, type of smoke, construction, the day, time of day, cars in the driveway, toys in the yard, mailbox overflowing, neighborhood), which plays into the implicit processing system and, in turn, influences the response.

This priming, procedural, or conditioning process requires less energy and can facilitate a shutdown of the neocortex, which allows other parts of the brain that would normally not be used simultaneously to begin to fire. This is what is sometimes called *time compression* or *time dilatation*, in which time seems to slow down. We all know that time is not slowing down. This is a result of the other areas of the brain working together while the neocortex is shut down, allowing us to see greater details in our circumstances than we would normally see. The ability to see these details in such vivid clarity gives us the impression that time is slowing down.[8]

As a company trains together, they develop an unspoken understanding. This unspoken understanding creates a momentum of its own. Using the example previously given in the explicit/implicit section, this company arrives at the fire scene. Everyone clearly understands the situation and their roles in this situation. The information they were given, from the drop of the alarm to the information updates over the radio, has been combined with the previous information gathered on the way to work (weather, time of year, holiday, pay day, weekend, etc.). They are rolling onto the scene, where rapid pattern recognition begins to galvanize their actions or bring about an awareness of something "different" that would alter their actions. The entire crew is on the same page, and very little communication is witnessed. It is all action. While the officer completes a 360, the rest of the company is busy with forcible entry, laying lines out for the attack, and "throwing" ladders. All responsibilities are being covered very quickly and with very little communication. Of course, some orders will be given because every fire presents its own set of circumstances, and other companies may arrive simultaneously, but the example of crew cohesiveness is just as applicable.

When all the egos and actions of the responders merge into a cohesive effort, control of the situations occurs. This cohesiveness happens when crews work together and share the same level of commitment, sacrifice, and focus. The company's commitment is developed in training and is polished during the mastery stages. Developing mastery is driven by risk because high consequences catch our attention and sharpen our focus. The results from an experienced crew are obvious through their consistent successes. Victories are not about luck; they are earned on the drill ground.

NOTES

1. "Yerkes-Dodson Law," *Wikipedia*, https://en.wikipedia.org/wiki/Yerkes%E2%80%93Dodson_law.
2. B. L. Fredrickson and T. Joiner, "Positive Emotions Trigger Upward Spirals toward Emotional Well-Being," *Psychological Science* 13 (2002): 172–175.
3. Carolyn H. Declerck, Christophe Boone, and Toko Kiyonari, "Oxytocin and Cooperation under Conditions of Uncertainty: The Modulating Role of Incentives and Social Information," *Hormones and Behavior* 57, no. 3 (March 1, 2010): 368–374.
4. Colette Bouchez, "Serotonin: 9 Questions and Answers," WebMD, https://www.webmd.com/depression/features/serotonin#1.
5. Mihaly Csikszentmihalyi, *Applications of Flow in Human Development and Education: The Collective Works of Mihaly Csikszentmihalyi* (Dordrecht, Netherlands: Springer, 2014).
6. Steven Kotler, *The Rise of Superman: Decoding the Science of Ultimate Human Performance* (London: Quercus Publishing, 2014).
7. Ibid.
8. Mihaly Csikszentmihalyi, *Flow: The Psychology of Optimal Experience* (New York: Harper Collins, 1990). A condensed version, "The 8 Elements of Flow," is available at http://www.flowskills.com/the-8-elements-of-flow.html.

Index

A

accountability, 114
   buddy system for, 51
   firefighters with, 43
   importance of, 51
adult learning
   five laws of, 30–33
   with training, 30
advanced life support (ALS), 52
age
   fear and, 16–17
   firefighters and, 16
   performance with, 16
anchor phrase
   development of, 104–105
   examples of, 104
   self-talk and, *104*
anxiety
   experience and, 21–22
   for firefighters, 14
   panic and, 16
Archilochus, 44
Asken, Michael, 99

B

bed nucleus of the stria terminalis (BNST), 19
belly breathing
   benefits of, 120
   as diaphragmatic breathing, 119
   importance of, 119–120
   technique of, 122
   in training, 122
Big Four
   for firefighters, 65
   for US Special Forces, 72
biometric tracking
   body core temperature in, *139*
   charts for, *138*
   example of, 139
   instructor with, 136
   technology for, 138
   training with, 136–139
   vitals for, 136
blackout exercise, 135
Blue Angels, *93*
Bowman, Bob, 77–78
box breathing, 65
   for fire fights, 63
   rookies and, 63
   square breathing as, 121
   for stress, 121
brain waves
   explicit system of, 141
   heart rate with, *140*
   implicit system of, 141
   role of, 140–141
bravery, 8
Brunacini, Al, 68
buddy system, 51

C

Carroll, Pete, 77
Cato, Kelley, xi
character
   bravery for, 8
   courage for, 8
   diligence for, 9
   experience and, 5
   for fire service, 5
   firefighters with, 3
   humility for, 6, 9
   leadership and, 6, 10–11
   loyalty for, 9–10
   mindfulness and, 124
   moral aspects of, 5
   self-assessment for, 17
   trust for, 7–8
   virtues and, 7–10
charter, 10
coaching
   failure-based training and, 32
   instructor to, 43–44
   learning events and, 32–33
communications, 52
confidence. *See also* SCBA confidence course
   achieving mastery and, 31–32
   law of primacy and, 31
   resiliency and, 136
   self-talk for, 106–107
constructive self-talk
   self control and, 99
   use of, 99–100
control measures, 51
coping statements, 106–107, 127

coping techniques
   motor memory and, 127
   repetition of, 126
courage
   for character, 8
   firefighters with, 8

D

diaphragmatic breathing, 119
diligence, 9
dirty breathing, 32
dopamine, 136
drilling
   in fire service, 35
   for firefighters, 33
   goals in mind with, 46
   with instructor, 50
   learning exercise and, 35
   objectives for, 50
   success-based training with, 40
   testing in, 33
   training and, 30
drone point of view, 86–87
dynamic imagery, 92–93, *94*

E

electroencephalograms (EEGs), 140
experience
   anxiety and, 21–22
   character and, 5
   fear with, 13
   for humility, 22
   participant knowledge and, 38
   situational awareness with, 66
exposure therapy
   fear and, 20
   post-traumatic stress disorder and, 20

F

failure-based training, 32
fear
   age and, 16–17
   exposure therapy and, 20
   in fire service, 13
   mental health and, 22
   in sympathetic nervous system, 13
   thinking traps and, 97–98
   training and, 13–16
fire call
   example of, 63–64
   trauma of, 75
   visibility during, 64

fire department
   firefighters in, 7
   location of, 37
   planning for, 49
   recruiters for, 67
   with training, 27
fire equipment, *85*
fire fighters
   communications and, 52
   during fires, 61
fire fights, 63
Fire Protection Association (NFPA), 37
fire science, 11
fire service
   career in, xi
   character for, 5
   danger in, 38, 49
   drilling in, 35
   fear in, 13
   instructor and, 135
   mental health for, *100*
   mental toughness for, 77–78
   with rapid intervention crew (RIC), 52
   rehabilitation in, 45
   required reading for, 27
   self control in, 65
   self-talk for, 101–103
   training for, xii, 14, 65
   virtues and, 10
firefighters
   with accountability, 43
   age and, 16
   anxiety for, 14
   background of, 107
   Big Four for, 65
   character for, 3, 9
   with courage, 8
   drilling for, 33
   in fire department, 7
   leadership for, 7
   learning for, 31–32
   mastery for, 143
   "Mayday" and, 66
   mental comprehension for, 3
   mental rehearsals for, *74*, 82, 94–95
   mindfulness for, 124–125
   pattern recognition for, 142
   physical exercise for, 130–131
   rehabilitation and, 45
   resiliency for, 81
   segmentation and, *112*
   self control for, 65
   with self-contained breathing

Index

apparatus, 89
  sleep for, 133
  stress for, 21, 124
  sympathetic nervous system and, 18
  technical skills for, 67
  training for, 15, 75
  visualization for, *83*
  wheel technique for, *123*
  Zephyr BioHarness for, 137
fireground, 32–33, 40, 45, 63, 81, 114
  lessons on, xi
  segmentation and, 111–112
firehouse, 35
fires
  calls for, 63
  example of, 61
  fire fighters during, 61
  hypervigilance against, xii
  winds and, xii
first-person (POV), 85
Fitbit, *80*
Fitzsimmons, Kim, 41
focus of activity, 36
freezing, 113–114

G

glucocorticoids (GCs), 19
goal setting
  objective goals and, 108
  outcome goals and, 108–109
  performance goals and, *109*
  process goals and, 110
  segmentation and, 107, 112
  subjective goals and, 108
  types of, 108–110
Gonzalez, Elvin, 28
Grossman, David, 82, 100

H

hazards, 51
hierarchy of needs
  components of, 6
  learning with, 30
  by Maslow, 6
high-performance skills, 69–70
hum technique, 120–121
humility
  for character, 6, 9
  experience for, 22

I

instructors
  with biometric tracking, 136

  to coaching, 43–44
  complaints and, 14
  drilling with, 50
  fear and, 13
  fire service and, 135
  learning from, 29
  SCBA confidence course and, 15
  for special operations forces, 77
  students and, 13–14, 32
  as training officers, 28
intermediate skills, 69

J

job performance requirements (JPRs), 68

L

large diameter hose (LDH), 37
law of primacy, 31
leadership
  character and, 6, 10–11
  development of, 6
  for firefighters, 7
learning. *See also* adult learning
  basics of, 29–33
  for firefighters, 31–32
  with hierarchy of needs, 30
  from instructors, 29
  psychology of, 29–30
  reviews for, 43
  threes laws of, 31
  in training, 29
learning events
  coaching and, 32–33
  with presentation, 42–45
learning exercise
  company activities for, 36
  drilling and, 35
  with large diameter hose, 37
Lesyk, Jack, 68
line-of-duty deaths (LODDs), xiii, 18
location, 49
  of fire department, 37
  training and, 37–38
logistics
  planning and, 37
  safety measures with, 38
loyalty, 9–10

M

Maslow, Abraham, 3
  hierarchy of needs by, 6
  training and, 30

mastery, 31–32
　for firefighters, 143
　over stress, 142
　training for, 142
"Mayday," 90
　emergencies and, 52
　firefighters and, 66
McGurk effect, 19
mental comprehension
　for firefighters, 3
　resiliency and, 66
mental health
　fear and, 22
　for fire service, 77–78, *100*
　resiliency and, 67
mental imagery, 82, 85
mental rehearsals, 69
　with dynamic imagery, 92–93
　firefighters with, *74*, 82, 94–95
　planning for, 88–92
　repeated practice of, 96
　timing of, 92
　types of, 81
　visualization and, 79–81, 83–84, 95
metabolic equivalents (METs), 130
mindfulness
　attention with, 125
　character and, 124
　for firefighters, 124–125
　neuroplasticity for, 125
　for self-contained breathing apparatus, *124*
moral
　character with, 5
　learning and, 29
motor memory, 106
　coping techniques and, 127
　muscle memory as, 127

N

Nash, Steve, 77
National Institute for Occupational Safety and Health (NIOSH), 18
Navy Seals
　psychological skills for, 72
　training for, 73
negative self-talk
　overcoming of, 125–126
　self-talk and, 126
　thinking traps and, 126
NFPA 1403: Standard on Live Fire Training Evolutions, 45, 53
NFPA 1410: Standard for Training on Emergency Scene Operations, 33, 37, 39
NFPA 1584: Standard on the Rehabilitation Process for Members During Emergency Operations and Training Exercises, 45
norepinephrine (NE), 19
nutrition
　diet for, 130
　poor nutrition and, 129
　sleep and, 132
　tips for, 129–130
　understanding of, 129–130

O

organization, 35
oxytocin, 136

P

participant knowledge, 38
PASS alarm, 90, *91*, 112
personal alert safety system (PASS), 88
personal protective equipment (PPE), 43, 50–51
physical exercise
　energy use in, 130
　for firefighters, 130–131
　metabolic equivalents in, 130
　types of, 131
planning
　contingencies in, 78
　for fire department, 49
　logistics and, 37
　for mental rehearsals, 88–92
　preparation and, 38–42
　time allotted for, 37
　training with, 35–36
poor nutrition
　diet for, 130
　nutrition and, 129
Posner, Michael, 28
postexercise debrief
　rehabilitation and, 46
　reviews and, 46
　in training, 46
post-traumatic stress disorder (PTSD)
　experience and, 21–22
　exposure therapy and, 20
　smell trigger of, 20
　stress and, 19
　virtues and, 3
Potterat, Eric, 71

Index

practice runs, 41–42, *95*
preparation
    planning and, 38–42
    training with, 38
presentation, 42–45

R

rapid intervention crew (RIC)
    fire service with, 52
    training for, 37
rapid intervention team (RIT), 90
rehabilitation
    in fire service, 45
    firefighters and, 45
    postexercise debrief and, 46
resiliency
    confidence and, 136
    for firefighters, 81
    mental comprehension and, 66
    mental health and, 67
    self control and, 66
    situational awareness and, 66–67
reviews
    for learning, 43
    postexercise debrief and, 46
    training with, 36, 51
Riley Emergency Breathing Technique, 120–121

S

safety measures, 38
safety plan
    example of, *54–57*
    features of, 49–53
    notes on, 53
SCBA confidence course
    complaints about, 14
    cycles of, 14–15
    foundation of, 15
    instructors and, 15
    students and, 15
    training with, 14
search and rescue (SAR), 66, 135
segmentation
    for crisis, 111–113
    firefighters and, *112*
    fireground and, 111–112
    goal setting and, 107, 112
self control
    constructive self-talk and, 99
    for firefighters, 65
    mastering of, 67

resiliency and, 66
self-talk and, 97
for US Special Forces (SF), 65
self management
    high-performance skills of, 69–70
    intermediate skills of, 69
    paradigm of, *68*
    performance skills of, 68
    psychological skills of, 68–69
    self-contained breathing apparatus and, 70
self-assessment, 17
self-awareness, 67–70
self-contained breathing apparatus (SCBA), 45
    firefighters with, *89*
    hum technique for, 120
    mindfulness for, *124*
    self management and, 70
    wheel technique for, 123
self-talk, 100. *See also* constructive self-talk
    anchor phrase and, *104*
    for confidence, 106–107
    fire crew and, *102*
    for fire service, 101–103
    on instructional, 102
    for motivational, 101
    negative self-talk and, 126
    proactive tool of, 105–106
    on reactionary, 103
    self control and, 97
    thinking traps and, 97–99
    on unrelated, 102
setting objectives, 39–41
situational awareness
    with experience, 66
    on internal, 66
    resiliency and, 66–67
    for US Special Forces, 67
    with training, 67
sleep
    for firefighters, 133
    health benefit of, 131–133
    nutrition and, 132
    sleep bank for, 131–132
    stages of, 132
SMART goals
    designing of, 111
    meaning of, 110
US Special Forces (SF)
    Big Four for, 72
    self control for, 65

situational awareness for, 67
Special Warfare Development Group and, 71
special operations forces (SOF), 73–74, 77
Staniszewski, Michal, 69
stress
   box breathing for, 121
   effects on brain, 18
   for firefighters, 21, 124
   mastery over, 142
   origins of, 18–21
   post-traumatic stress disorder and, 19
   sight/sound and, 19
   smell and, 20
students
   instructors and, 32
   SCBA confidence course and, 15
success-based training
   with drilling, 40
   possibility of, 40
   setting objectives and, 39–41
surgeons, 71
sympathetic nervous system (SNS)
   breathing and, 75
   evolutionary tool of, 17
   extreme stress on, xiii
   fear in, 13
   firefighters and, 18
   hyperarousal of, 119

T

temperance, 9
testing
   in drilling, 33
   purposes of, 33
thinking traps
   fear and, 97–98
   negative self-talk and, 126
   overcoming of, 98
   self-talk and, 97–99
third-person point of view, *87*, 88
training
   for academies, 82
   adult learning with, 30
   belly breathing in, 122
   biometric tracking in, 137
   center for, 28
   description of, 50
   drilling and, 30
   fear and, 13–16
   fire department with, 27
   fire service with, xii, 14, 65
   for firefighters, 15, 75
   gain confidence with, 13
   key messages in, 41
   learning in, 29
   location and, 37–38
   for Navy Seals, 73
   officers for, 28
   organization for, 35
   with planning, 35–36
   postexercise debrief in, 46
   with practice runs, 41–42
   with preparation, 38
   for rapid intervention crew, 37
   reviews in, 36, 51
   with SCBA confidence course, 14
   setting objectives and, 39
   situational awareness with, 67
   successful steps to, 29–30
   thriving with, 114
   weapons of mass destruction and, 27
Trujillo, Gus, xi, xii
trust, 7–8

U

US Fire Administration (USFA), 18
U.S. Firefighter Deaths Related to Training, 2001–2010, 53

V

vent, enter, search (VES), 66
virtues
   character and, 7–10
   fire service and, 10
   of particular, 6
   post-traumatic stress disorder and, 3
visualization
   for firefighters, *83*
   mental rehearsals with, 79–81, 83–84, 95

W

weapons of mass destruction (WMD), 27
wheel technique
   breathing with, 122–123
   for firefighters, *123*
   for self-contained breathing apparatus, 123
winds, xii
Wright, John, xi, xii

Zephyr BioHarness, 79, *80*, 137

About the Authors

BOB CARPENTER

Bob Carpenter is a 38-year veteran of the fire service, having recently retired from Miami-Dade Fire Rescue. His career spans volunteer fire service and career fire service as a member of a combination department. Bob served for 30 years with Miami-Dade (FL) Fire Rescue (MDFR).

As a captain, Bob was the bureau officer in charge of recruit training. He served as the training captain for the North Operations Division for 10 years, during which time he developed and deployed numerous training initiatives, ranging from company-level to department-wide projects.

Bob is the author of several training articles in internationally published trade magazines. In addition, he is a co-lead instructor with Tactical Resiliency Training, LLC, specializing in tactical resiliency training and officer/instructor development.

Since 2007, Bob has presented workshops and classroom sessions at the Fire Department Instructor's Conference (FDIC) International in Indianapolis, IN, speaking on the subjects of training and officer development. Bob's "Drill Development: Four Steps to Success" workshop has been part of the State of New Jersey's Department of Fire Safety Instructor continued contact curriculum through Kean University at more than 80% of the state's fire academies. His "Drill Development" workshop has been a core part of the mandatory Officer Development Academy at Palm Beach County (FL) Fire Rescue for three years. Bob Carpenter can be contacted at carpenter2156@bellsouth.net.

DAVE GILLESPIE

Dave Gillespie is a 25-year veteran with the Peterborough Fire Service in Canada. He has worked as firefighter, chief training officer, and incident safety officer, and he is currently on the trucks.

Dave has spoken at FDIC and authored articles for *Fire Engineering* and *Firefighting in Canada*, including "Due Diligence" and "Innovative Simulation Training in Acquired Structures." He is an adjunct professor at Fleming College Firefighting Program and teaches Forcible Entry, HazMat, and Fire Ground Operations courses.

A provincially certified trainer and department water rescue instructor for 20 years, he has provided training in fire-rescue operations, swift water and ice rescue, hazmat, air management, and other various subjects across the country.

Dave developed Ontario Fire College's Water/Ice Rescue Program and has served as a stunt safety coordinator, consulted with fire departments, and provided courtroom testimony on fire service's "industry best practices."

In 2011, Dave received the Chief's Excellence in Service Award for the development of YouTube video for the Swim to Survive program. With 800 Grade 3 students participating each year, more than 6,000 children have graduated from the joint-services/aquatic program.

From personal experience, Dave lived the stress response and has seen first-hand "how people don't rise to the occasion, they default to their level of training." That experience influenced his research and development of his training philosophy to help firefighters raise their performance, involving simulations, stress inoculation, and stress-induced practices, along with resiliency tactics.

Dave is a partner in Tactical Resiliency Training LLC, and serves as critical incident stress (CIS) debriefer. He is a trainer in "Road to Mental Readiness and Suicide Intervention" and speaks with corporate groups on high performance and resiliency.

He applies the training and resiliency principles outlined in this book with his family and at work on a daily basis.

RIC JORGE

Ric Jorge completed a 24-year career as a firefighter with Palm Beach County Fire Rescue, a district with more than 50 firehouses covering 1.2 million citizens over 1,762 square land miles. As a national and state-certified instructor teaching in his department for more than 14 years, Ric developed and delivered several department-wide trainings.

Ric has delivered lectures and Hands-On Training (H.O.T.) classes nationally and internationally (Ecuador Fire Academy, S.A.), as well as teaching at several well-known fire conferences and training academies annually across the country. These classes have been presented at the FDIC International (Indiana), Fort Lauderdale Fire Expo, Fire Rescue East (Daytona), Orlando Fire Conference (Orlando, FL), Metro Atlanta Fire Conference (Atlanta, GA), 50th State Fire Conference (Oahu, Hawaii), Florida State Fire Academy, Firehouse Expo (Baltimore and Nashville), Great Florida Fire School, Texas Association of Fire Educators Conference, Berks County Fire Symposium (PA), Command Officer Boot Camp (Pensacola, FL), Firemanship Conference (Portland, OR), NE Fire Summit (MA), TEAM Conference (Spokane, WA), and the Company Officer Leadership Conference (LSU/Fire & Emergency Training Institute).

Ric has developed several courses, two of which have drawn national interest: "Impact This: Forcing Impact-Resistant Openings" and "The Monster/Courage Within: Developing and Teaching Resiliency."

Ric has authored chapters in books, including *Fire Engineering's Handbook for Firefighter I and II* and Dave Dodson and John Mittendorf's *The Art of Reading Buildings*. He has also authored articles for publication in magazines (*Fire Engineering* and *1st Responder*), newsletters ("Fire Department Training Network"), and online blogs ("Backstep Firefighter").

Ric Jorge can be contacted at surfdogs4@yahoo.com. Some of his work can be found on FaceBook under "Tactical Resiliency Training," as well as on YouTube.

Dave lives a busy life with Monica, his wife of 22 years, and his sons, Brock and Dawson, doing runs, remote canoe trips, swimming whitewater, and any adventure that makes them laugh.